FRC Design Handbook

# Macro Design: Preparing for the Game Drop

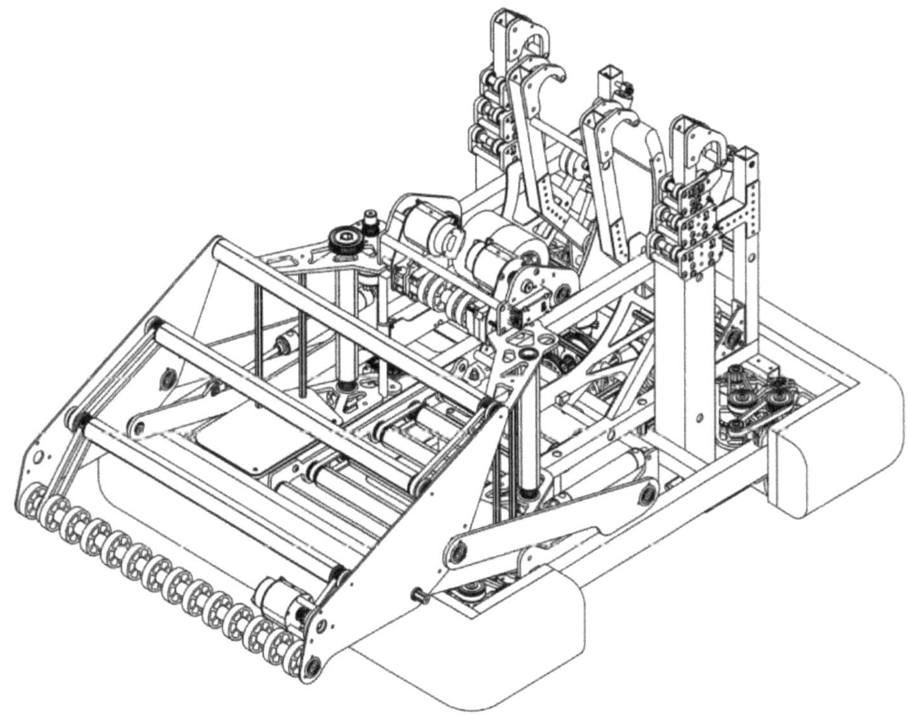

Avery W. Fiore

**THE AUTHOR**

**Avery W. Fiore** entered the world of robotics competition aged 10. He has led teams and won numerous awards in FIRST® LEGO® League, VEX® IQ Competition, VEX® Robotics Competition and FIRST® Robotics Competition. In 2020, when the world came to a standstill due to COVID-19, he picked up CAD via a single Youtube video. With a bit of imagination, he designed a fully functioning robot that went on to win the Excellence Award at VEX Worlds 2021. Always passionate about sharing his love for robotics, he has invested over 1,000 hours coaching junior members. He wrote Macro Design: Preparing for the Game Drop to complement CAD classes he conducts for students. Avery was the FIRST® Robotics Competition Captain for Team 4817 at the Singapore American School from 2022 to 2024. He completed a two-year stint with the Singapore Armed Forces and is currently enrolled in Electrical and Computer Engineering at Cornell University.



Updated printing 2026.1 adds Subchapter 2.3 Omni Drivebases.
Amendment packet for earlier printings: www.frc-handbook.com/update2026_1

Published by Fiori Books, an Imprint of Pivotal Axis Pte Ltd.



ISBN 979-8-9997828-0-9 (Paperback)
ISBN 979-8-9997828-1-6 (Hardcover)

Printed in U.S.A.

## FOREWORD

The FIRST® Robotics Competition (FRC) is a prestigious and longstanding robotics competition for high school students, drawing young minds with their passion for robotics. Despite its widespread popularity, many students encounter challenges in commencing their FRC journey due to the scarcity of beginner-friendly resources. The majority must rely on fragmented online materials or the limited expertise of teachers and upperclassmen.

This book provides an exceptional introduction to robot design employing CAD, specifically tailored for FRC. It elucidates the fundamental components of an FRC robot—including the drivebase, intake, and launcher systems—and demonstrates how these components are applied to address the challenges encountered in past competitions. Beyond CAD design, the book covers practical strategies for robot planning and operation. I firmly believe this book will serve as an indispensable guide for any student embarking on their FRC journey and an invaluable resource for mentors. It has the power to transform curiosity into capability and help students take their first confident steps into the world of robotics.

Dr. Seok Woo Lee
Associate Professor
School of Electrical and Electronic Engineering
Nanyang Technological University, Singapore

**CREDITS: IMAGE & CAD CONTRIBUTIONS**

I would like to express my sincere gratitude to all the teams who generously granted me permission to use their images and content. Your support and openness have been instrumental in bringing this project to life. Each of you have contributed not only valuable visual material but also inspiration through your creativity, innovation, and dedication. This spirit of collaboration reflects the very heart of the FIRST® Robotics Competition—where sharing knowledge, resources, and encouragement helps strengthen the entire community. I am truly thankful for the opportunity to showcase your work with proper credit and appreciation.

The teams whose robots and mechanisms are featured in this book are listed by numerical order:

- Team 118: The Robonauts
- Team 254: The Cheesy Poofs
- Team 971: Spartan Robotics
- Team 1986: Team Titanium
- Team 2910: Jack in the Bot
- Team 4613: Barker Redbacks
- Team 4817: One Degree North
- Team 5987: Galaxia in memory of David Zohar

Team 4817: One Degree North is the FRC team that I had the opportunity to flourish in. I would like to thank my fellow team members over the years from the CAD subteam whose contributions are included in this book: Ming Jin Yong, Ryan Lee, Emilio Orcullo, Adam Brest, Siddhant Gupta, Aryan Pershad, Daniel Xu, Munira Takalkar, Doy Yun, Yaahvei Krishna, Uddalak Sarkar, Anthony Park, Brian Chen, and Aarya Mishra.

Additionally, robots from a couple of RI3D teams are featured:

- UMN Robotics with credit to Katriana Trinh, Jace Flansburg, and Alex Tong
- Capital Robotics on the FUN Robotics Network

This book would not have been possible without the rich trove of CAD renderings from each and every designer who spent significant hours tinkering and refining individual mechanisms. Thank you.

## PERSONAL ACKNOWLEDGEMENT

The opportunity for me to flourish as a CAD designer would not have been possible without the foundational support of the Singapore American School's robotics program. My journey began with the FIRST® LEGO® League (FLL), where I was introduced to the rigor and joy of competitive robotics. Mr. Dan Gach, who founded the FLL program at SAS, cultivated an environment that was remarkably conducive to learning. It was in that space that I first encountered the challenges of collaboration—resolving disagreements, advocating for ideas, and navigating team dynamics. These lessons were not always easy, but they were formative. Looking back, it was a magical time—one in which I began to mature, not only as a budding engineer but as a person.

My first exposure to formal engineering concepts came with the VEX® IQ Competition (VIQC). It was here that I met my primary robotics mentor—someone who shaped my approach to design and instilled in me values that continue to guide me today. He taught me the power of giving back to the community and helped build my self-belief—the kind that makes difficult goals feel attainable. Remarkably, at the time, he was just a high school student. He had founded the VIQC program at SAS and yet found time to mentor younger students like me. To the high school students reading this: never underestimate the influence you have on those who follow in your footsteps. You are their role models. The way you lead shapes the way they dream. I was fortunate to have such a role model in Rohit Narayanan.

Rohit once told me,

"Everything you need to learn in life, you will learn in competitive robotics."

I've found that to be increasingly true as I progressed through more advanced competitions. In my first year of high school, I seized the opportunity to lead the first ever SAS team to VRC Worlds where we won the Excellence Award. That journey taught me invaluable life skills— like perseverance, teamwork under pressure, and how to turn failure into iteration. From my sophomore year onward, I transitioned to FRC, where I was mentored by captains Ryan Lee, Ming Jin Yong, and Tanisha Mehta. They showed me how to effectively lead a complex, interdisciplinary organization like an FRC team.

It was during this period that I began developing CAD and design training resources to help junior members within Team 4817. That internal effort grew into the foundation of the FRC Handbook series. In my senior year, I had the privilege of co-leading the team alongside Yohan Sakurai and Marcus Chin. Together, we navigated one of the most successful seasons in Team 4817's history. I could not have done it without them.

That season was the result of shared vision, relentless effort, and deep trust among teammates. I'm especially grateful for my graduating cohort—Jake Lee, Frank Xie, Benjamin Turney, Ella Chi, Asia Marcos, Alice Yang and Lauren Lim—who shaped the team's culture through their leadership, collaboration, and commitment across the years we spent together. Whether they joined early or later, each left a meaningful mark on our journey.

I am also deeply grateful to the teacher sponsors who kept the high school robotics lab open long after hours—giving us the time and space to meet deadlines and power through crunchtimes. Mr. Bart Millar and Ms. Meredith White were the first sponsors of the High School Robotics Club. They met students where they were—granting us the autonomy to make decisions and the responsibility to own the outcomes. Mr. Paul Booth and Mr. James Harvey carried that legacy forward, sustaining the student-led spirit of the program. The tradition of supporting student passion extended to middle school as well, where Mr. Stan Richards corralled younger students to stay the course. Their steady presence has been instrumental in nurturing a culture of curiosity and excellence.

None of this would have been possible without funding and institutional support. I am thankful to the parents and donors who believed in a hands-on, student-driven model of STEM education. I also extend my gratitude to SAS superintendents Dr. Chip Kimball (2012–2019) and Mr. Tom Boasberg (current), who consistently listened and supported efforts to expand access to STEM and robotics at SAS.

Most recently, as I prepared this handbook, I had the opportunity to test its content with a few younger students. Thank you, Alex Ng, Ian Lee, and John Cruse—for your thoughtful feedback, honest critiques, and willingness to explore new material with me. It was your curiosity and engagement that reignited my passion for teaching engineering—and ultimately led to the publication of this book.

# Contents

# 1. Overview

## 1.1 Introduction

The FIRST® Robotics Competition (FRC) is arguably the apex of high school robotics competitions. The level of complexity and variability in the games makes it a challenge even for professional engineers to coach FRC teams. This makes it all the more rewarding and fulfilling when students are able to come together and succeed at FRC events. Over 4000 FRC teams compete around the world annually to first past district/regional level events, and then to duke it out at the FRC World Championships. It is the hope of many teams and individuals each year to further refine their engineering skills to progress further in the competition and to continuously grow in learning.

While there are a few ways to succeed in FRC, the primary way centers around the robot an FRC team makes. The beginning of the robot build process starts with design.

In the FRC robotics season some would say that Kickoff is the most exciting day. Kickoff is when the robot game is revealed. This usually occurs on the first Saturday/Sunday (depending on timezone) of the New Year. FRC teams around the world spring into overdrive. Within the first 24 hours of the game reveal, teams will gather, analyze the new FRC game, and brainstorm a rough idea of what the team's design will be. From there, most teams will start to CAD their design.

So how does a team go about designing an FRC robot? A team must be able to imagine the robot before it is built. Some teams leave this job to the most talented members as top designers will often describe an ability to mentally visualize and manipulate components in 3D space to assemble a complete robot concept. However, full participation by all members during the Kickoff is much more effective at engaging the full team from the get-go.

Why are some members better than others in the brainstorming session at Kickoff? In order to have the creativity to imagine a new robot design, a member must first have the base wealth of knowledge of good FRC designs of the past.

This handbook prepares FRC members for Kickoff day. The bulk of the handbook covers content knowledge of the different mechanisms that are most frequently used, with the final chapter about the Kickoff Design Discussion process. While the rest of this book series is targeted at CAD sub-team members on how to design FRC robots, this particular volume is written for all members in an FRC team so that they can effectively contribute to the Kickoff Discussion.

This handbook provides many past examples of FRC robots to inspire new designs. What gives individual teams an edge in the competition is the ability to creatively implement general concepts of mechanisms to solve problems uniquely. When that is done properly it can lead to a more effective robot.

## 1.2 How to Use This Handbook

This handbook is meant to be more than a collection of ideas—it is a resource that can be adapted to the needs of different readers and teams. Depending on your role and context, you may find one of the following approaches most useful:

1. **Individual Reading and Self-Study**
   Students who want to strengthen their design intuition can read through this book at their own pace. The examples and explanations are written to be accessible for both new and experienced team members.

2. **An 11-Week Team Curriculum**
   Teams may choose to adopt a structured learning path, covering one chapter per week across an eleven-week period. In this model, members read individually and then gather once a week to discuss the concepts as a team. This rhythm not only builds technical knowledge but also fosters communication skills that are vital during Kickoff.

3. **Preliminary Training for CAD Sub-Team Members**
   For students joining the CAD team—or those returning who need a refresher—this handbook serves as a starting point before diving into software training. Reading through selected chapters equips new members with the vocabulary and design patterns they will soon translate into CAD models.

4. **Reference for Specific Mechanisms**
   When preparing for a competition or reviewing design options, teams can use this book as a quick reference guide. Each chapter highlights the strengths, weaknesses, and trade-offs of common mechanisms, allowing for informed decisions under time pressure.

5. **Mentor Resource**
   Mentors may also find this book valuable as a supplement to their own training materials. It provides shared language and examples that can support lesson planning, discussions, or design reviews.

In short, this handbook is meant to meet you where you are—whether you are a student exploring design concepts for the first time, a team building shared understanding through structured study, or a mentor guiding students toward better engineering practice.

## 1.3 Mechanisms

There are many, many different mechanisms to choose from when designing a robot, with variations on each. This handbook can only give an overview of the most widely used mechanisms. Chapter 2 covers drivebases; Chapter 3 intakes; Chapter 4 indexers and storage systems; Chapter 5 flywheels; Chapter 6 catapults; Chapter 7 arms; Chapter 8 lifts/elevators; Chapter 9 multi-axis arms; Chapter 10 claws; Chapter 11 motor sharing techniques. Each chapter is written to be concise enough to refresh your memory during competition preparation, while still offering depth for structured study.

## 1.4 Kickoff Design Process

The final chapter in this volume covers the general Kickoff design process an FRC team goes through as soon as the new season's game is announced. It is at the Kickoff Design meeting that all team members – regardless of specialisation in mechanical, electronics, programming, CAD, outreach, etc. – come together to brainstorm solutions to the game challenge. At this time, each member should have a good understanding of the most common mechanisms used in FRC competitions.

## 1.5 Online Resource References

Further online resources are provided at www.FRC-handbook.com. Specific mechanisms are explained by type, but most mechanisms featured are part of a functioning robot. To understand how the mechanisms work in the context of a full fledged design, please refer to the online links at each relevant section. Each diagram includes a reference number that can be found on the website and typically includes a 3D render of the full robot. Additionally, a few animated diagrams and links to videos can be found on the website. It will be helpful to bookmark this website as extensive references are made to it throughout this book.

Note that the example files are taken from Team 4817 unless otherwise stated.

# 2. Drivebases

## 2.1 Introduction

The drivebase is the base platform of the robot that enables the robot to move on the field. Many different drivebase designs have been developed over the years, with some outclassing others. This chapter covers the most widely used FRC drivebases, but not all are considered suitable in modern FRC games.

Drivebase design involves 3 main considerations: complexity, maneuverability, and defensibility. Simpler is generally better. Being able to move your robot in multiple directions can allow you to outmaneuver your opponents. In the event your robot needs to play defense in the alliance or simply to fend off contacts with opposing robots, you want your robot to not be easily pushed around.

## 2.2 Tank Drivebases

A standard tank drivebase uses 4 wheels, in 2 sets of parallel wheel configurations. In the example below, all 4 wheels used are traction wheels.

Diag. 2.1: Tank Drivebase Ex. Isometric View; Ref. 1

While this makes a decent drivebase for moving backwards and forwards and even to push against other robots, it does not turn well. When attempting to turn, this drivebase produces a skidding sound as the motors struggle to overcome the friction between the wheels and the floor since this involves some lateral movement.

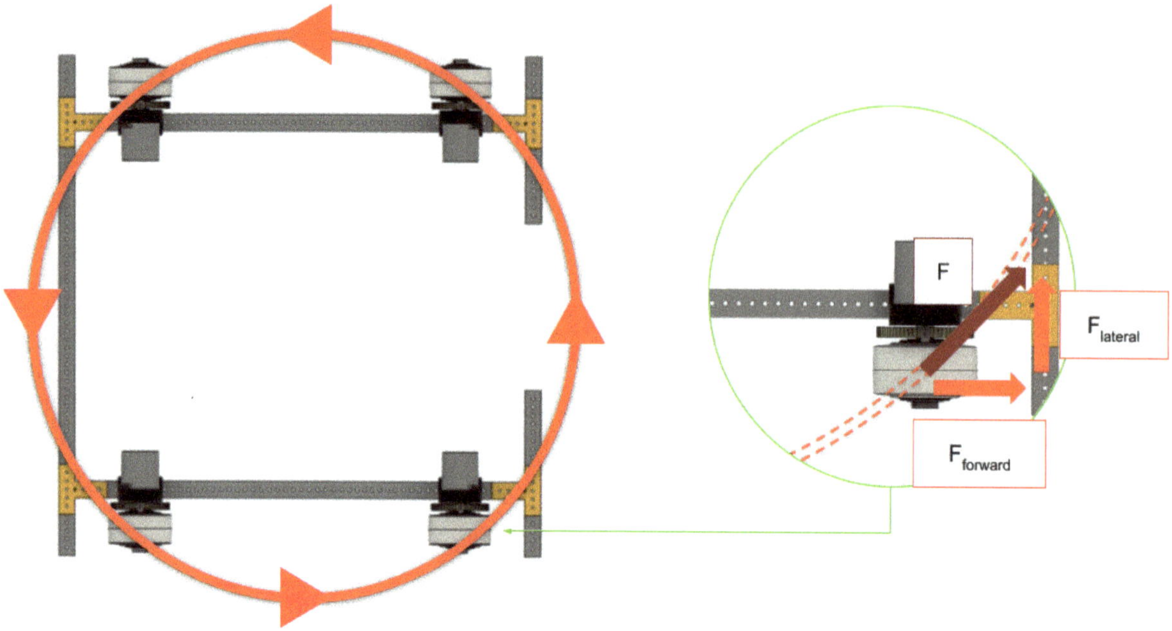


Diag. 2.2: Tank Drivebase Ex. Turning Circle; Ref. 1

As shown in the diagram above, given that the 4 wheels are in a squarish arrangement, the turning circle of the drivebase that follows is shown on the left. Zooming in to the bottom right wheel, it is shown that the force vector the motor attempts to exert on the wheel follows a roughly 45 degree upward angle. This means while some of this force is going into moving the wheel forward, roughly half is going into moving the traction wheel sideways. This problem applies to all 4 wheels in a similar way. This creates the skidding problem and stops the drivebase from being able to turn effectively. This problem can be overcome by switching two of the wheels to omni wheels.

If at least half of the wheels are switched to omni wheels, the resultant drivebase is known as an omni drivebase.

## 2.3 Omni Drivebases

An omni drivebase comes in 2 forms.

The first type of omni drivebase uses 4 omni-wheels on a tank drivebase chassis. It is very maneuverable, however, the robot has zero pushing and defence capability. An opposing robot can easily ram the robot from the side, causing the drivebase to strafe.

A second and more effective type of omni drivebase uses 2 traction wheels and 2 omni wheels. The omni wheels allow the robot to turn smoothly, while the traction wheels prevent the robot from being pushed as easily from the side. In the rare instance where a force is exerted from the side near the omni wheels, as that end of the robot has less lateral grip, the robot would pivot. (The pivot point would be between the robot's 2 traction wheels.)

An example of such an omni drivebase is shown here of a robot made by Team 118. See https://www.frc-handbook.com/ref/13.

Diag. 2.3: Omni Drivebase Ex. Bottom View; Ref. 13

This omni drivebase also has a pair of additional tracking wheels. These wheels rest lightly on the ground and do not affect the robot's turning capabilities and have no propulsion. They are linked to an encoder to accurately measure the robot's position and movement on the game field. The main reason for using tracking wheels is because in an omni drivebase, using encoder readings on the drivebase's main wheels will lead to inconsistency, drift, or inaccurate position readings due to wheel slippage. The additional tracking wheels get around this problem.

While the omni drivebase is an improvement over the tank drivebase, there is a better way to solve the skidding problem of traction wheels—the West Coast Drivebase.

## 2.4 West Coast Drivebases

A west coast drivebase (WCD) uses 6 to 8 wheels arranged in 2 parallel lines. The middle 1-2 wheels on each side are lowered, usually by ⅛" from the outer wheels.

Diagram 2.4 shows an 8-wheel example of a WCD.

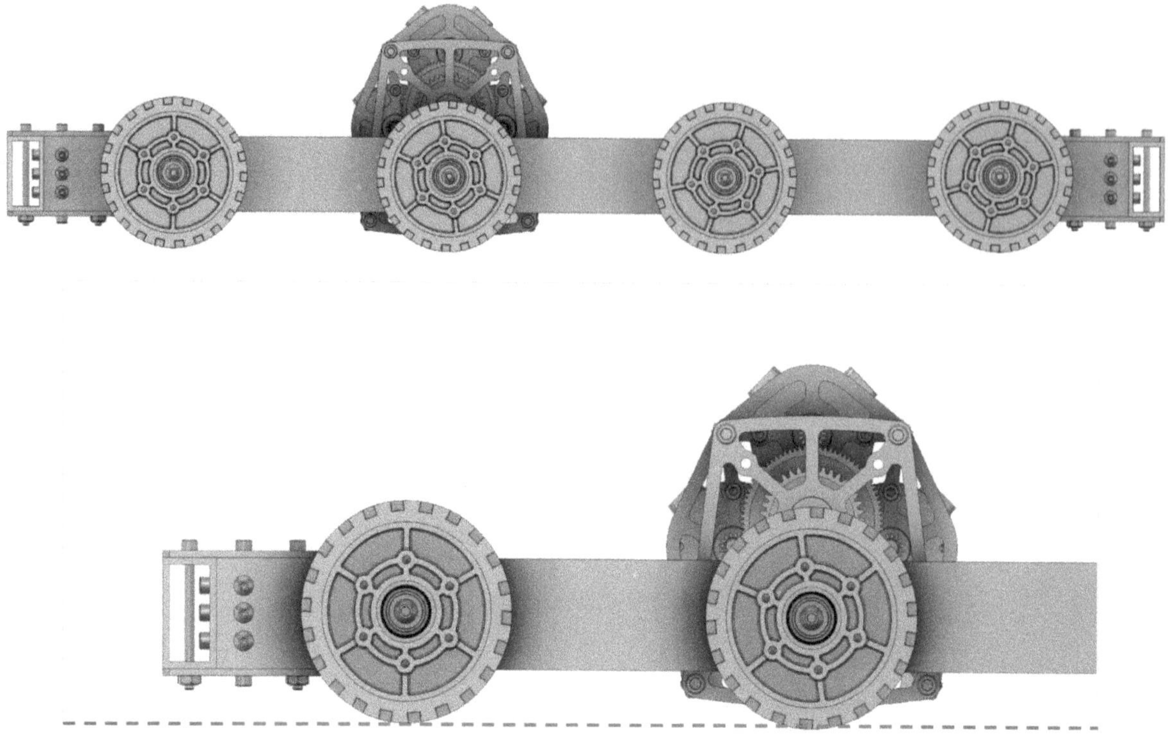

Diag. 2.4: 8-Wheel WCD Ex. Wheel Drop; Ref. 2

Below is a 6 wheel example of a WCD.

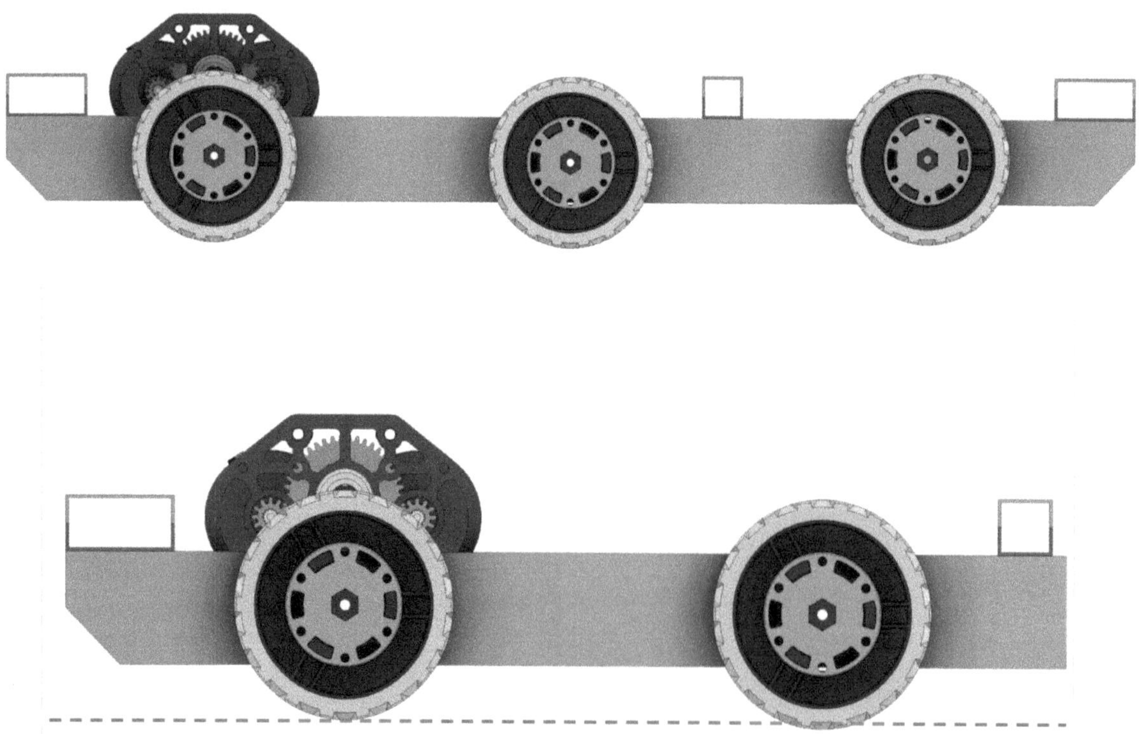

Diag. 2.5: 6-Wheel WCD Ex. Wheel Drop; Ref. 3

184

It's a bit hard to see in both examples as ⅛" is not that much. However, the center wheel(s) are indeed slightly lower than the outer wheels.

The theory behind why a WCD can turn properly despite using all traction wheels hinges on how the turning circle interacts with the wheels that are contacting the ground. In the case of an 8 wheel WCD, the robot is usually sitting on the center 4 wheels, with the outer wheels only coming into contact with the ground when bumping into opposing robots or in sudden changes in direction. In the case of a 6 wheel WCD, the robot either tips forward or backward still resting on only 4 wheels at any given time.

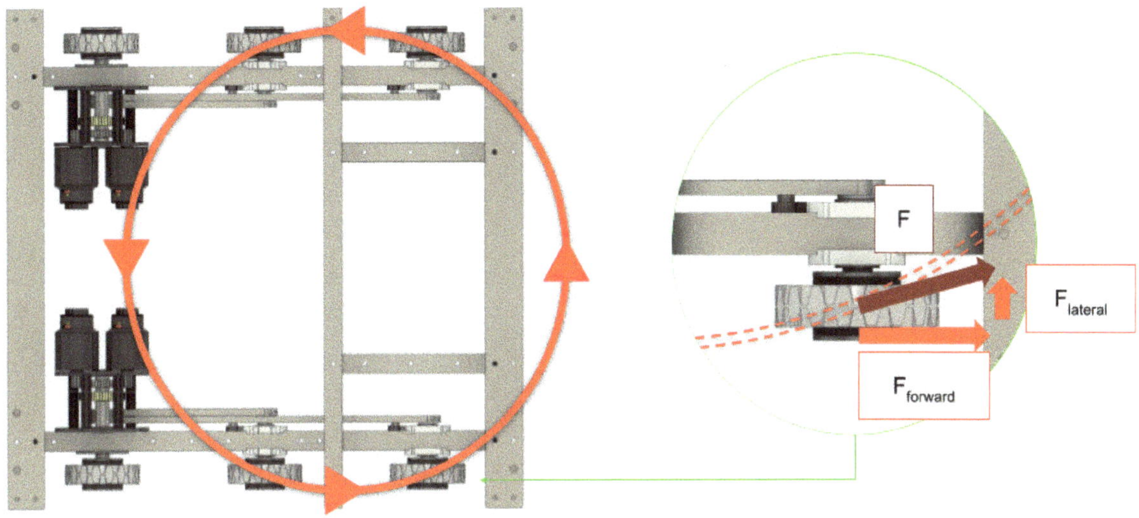


Diag. 2.6: 6-Wheel WCD Ex. Turning Circle; Ref. 3

As shown in the diagram above, even while turning, a WCD is mostly applying a force forward with not that much force shifting the wheel sideways. This makes it in essence a Tank Drivebase whilst in operation, (as there are only 4 wheels actually contacting the ground at any given time), that has a significantly greater width than length.

This allows a WCD to turn properly, and as only traction wheels are used, a WCD keeps all of the advantages that comes with a 4-Wheel Tank Drivebase. A WCD can be an effective design for pushing opposing robots.

In each set of wheels on either side of a WCD, there is a gearbox powering all of the wheels in the set. The wheels use some kind of power transmission to link with the gearbox. Each gearbox of a WCD uses anywhere from 2-4 motors (meaning 4-8 motors total), depending on how much torque and speed is required in the design.

## 2.5 H-Drivebases

A H-drivebase takes the layout of an all omni wheel drivebase and adds an extra wheel perpendicular to the other 4 wheels in the center of the drivebase. This allows the drivebase to move laterally.

Given that all of the wheels are omni wheels, opposing robots can easily push a robot with this type of drivebase around, particularly from the side where there is often only 1 motor powering the drivebase. Even when facing a tank drivebase head on with the same number of motors engaged, as omni wheels have less traction than traction wheels, a H drivebase will lose in a pushing match.

In robotics, there are often trade-offs between different design options. The tradeoff here is greater maneuverability for significantly weaker defensive capabilities. This tradeoff is considered costly enough that a H-drivebase is shunned by most teams.

Regardless, here is an example of what a H-drivebase could look like.

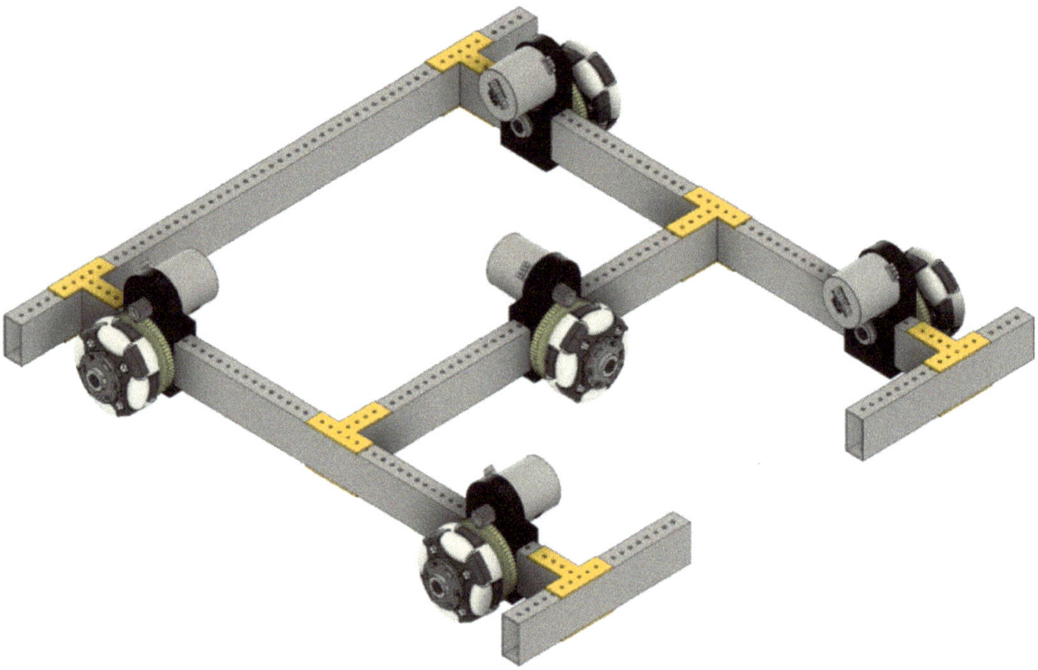

Diag. 2.7: H Drivebase Ex. Isometric View; Ref. 4

As seen, there is typically a single motor powering the lateral movement wheel, though a gearbox with 2 motors could be used. The other 4 wheels can follow the power transmission set up for any standard tank drivebase. The first option is where each wheel is powered directly by one motor (as shown here). The second option is where each side is powered by a gearbox of motors that are linked to the same side wheels of the drivebase.

## 2.6 Mecanum Drivebases

A mecanum drivebase uses mecanum wheels to achieve the mobility of a H-drivebase without the need for a lateral wheel, motor, and gearbox. It is the design of mecanum wheels, (with many mini wheels slanted diagonally), that makes it possible for this to be achieved. An FRC mecanum drivebase would look something like what is shown in diagram 2.8.

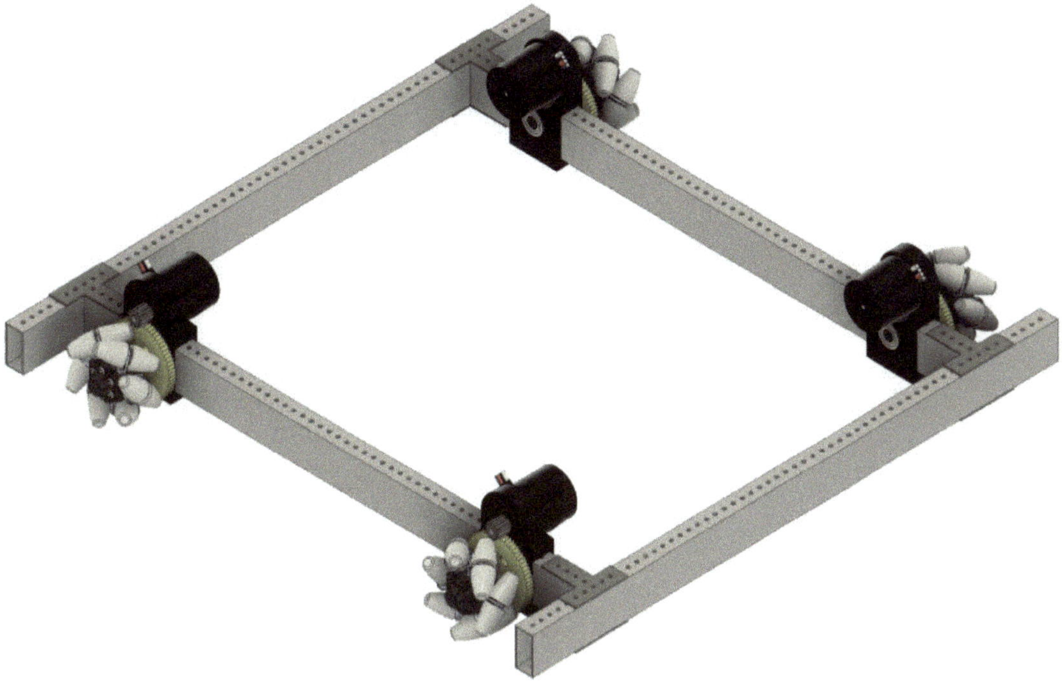

Diag. 2.8: Mecanum Ex. Drivebase Isometric View; Ref. 5

A big difference between a mecanum drivebase and other drivebases is that each wheel must be individually motorized. Otherwise, lateral movement cannot be achieved. This is because the front and back wheels have to spin in opposite directions to achieve lateral movement. An example of a VRC mecanum drivebase in motion can be seen at https://www.frc-handbook.com/ref/6.

Given the lateral capabilities of a mecanum drivebase, this drivebase suffers from similar issues to a H-drivebase when it comes to defense. Perhaps such problems theoretically are less as all motors can be engaged when driving forward and laterally, but in reality, mecanum wheels have even weaker traction than omni wheels, making defense with this type of drivebase difficult.

It should be noted that given the direction of the mini wheels on the mecanum wheels, turning is very smooth with this drivebase.

## 2.7 Octocanum Drivebases

An octocanum drivebase has 4 modules of wheels that allow the drivebase to switch between a traction mode and a mecanum mode. This allows for the drivebase to maintain the full maneuverability of a mecanum drivebase, while having a tank drivebase to fall back on for defensive maneuvers. Such a drivebase is considered significantly more complicated than any other drivebase discussed so far.

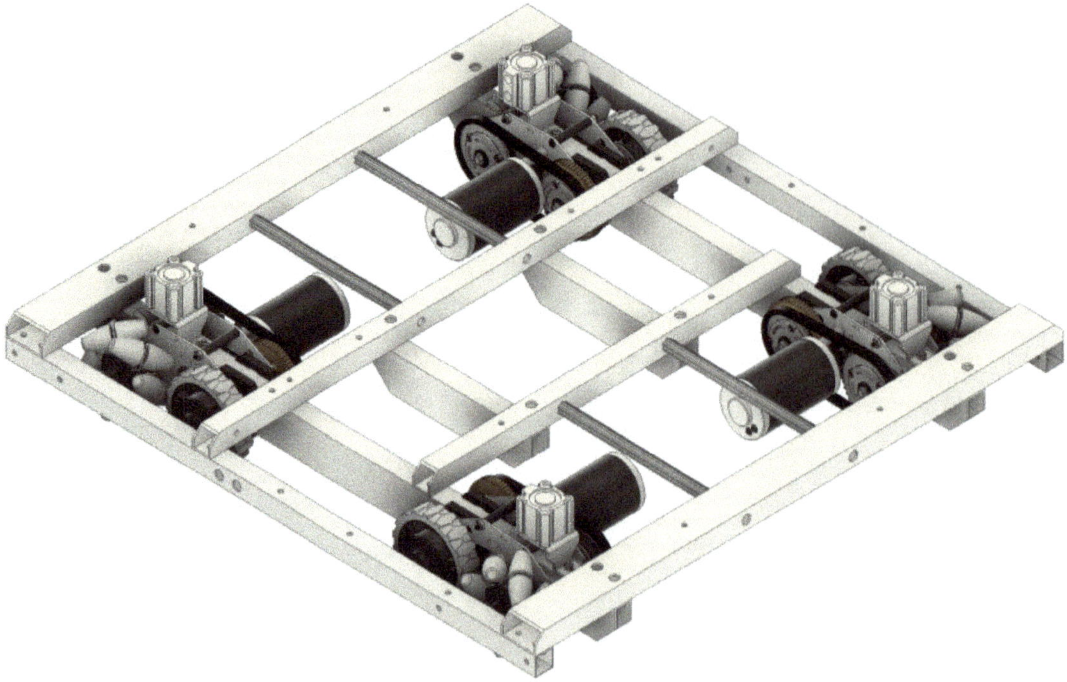

Diag. 2.9: Octocanum Ex. Drivebase Isometric View; Ref. 7

Shown above is an octocanum drivebase. Let's zoom in to see how one of these individual modules work. (Refer to https://www.frc-handbook.com/ref/7 for more information on the project and full robot.)

Diag. 2.10: Octocanum Ex. Power Transmission; Ref. 7

In diagram 2.10, the motor is hidden from view. The gear in the middle, the driving pinion, is powered by the motor. This in turn drives the gear beside it and creates the necessary gear reduction to enable the drivebase to function effectively. The mecanum wheel and traction wheel are both powered by the same motor via a chain that wraps around sprockets on both shafts. In summary, both wheels spin in the same direction whenever the motor spins.

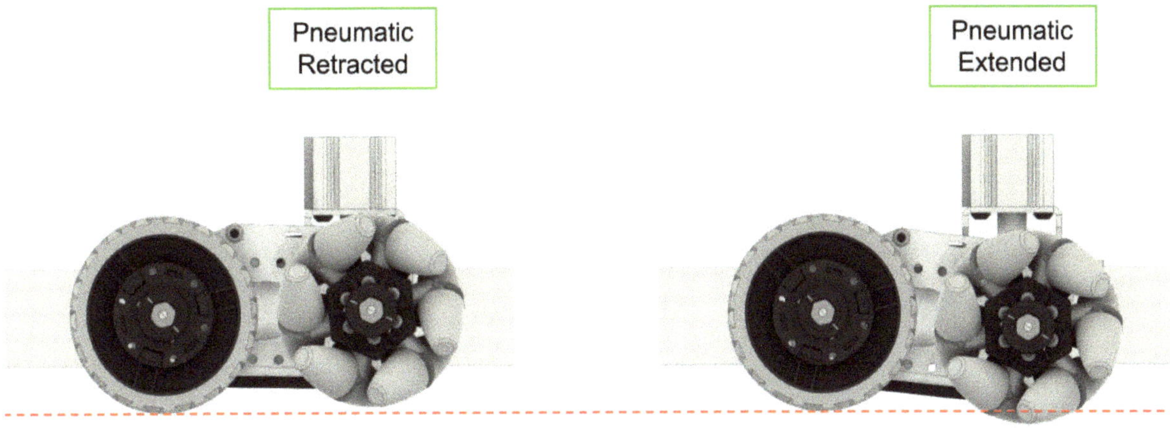


Diag. 2.11: Octocanum Ex. Pneumatic Positions; Ref. 7

A pneumatic controls which wheel is contacting the ground. When the piston is retracted, the traction wheel contacts the ground. When the pneumatic is extended, the mecanum wheel contacts the ground. This is what allows the octocanum drivebase to switch between traction and mecanum modes.

Note that in a good octocanum drivebase design, the traction wheels are positioned closer to the center of the robot. This is so that the effective drivebase when in traction mode is relatively wide to how long it is. Such proportions allow the drivebase to turn in traction mode.

## 2.8 Swerve Drivebases

A swerve drivebase approaches the problem of maintaining traction while keeping high maneuverability a bit differently. Each traction wheel is in a wheel module powered by 2 motors. 1 motor drives the wheel itself, while the other motor changes the direction of the wheel.

Diag. 2.12: Swerve Drivebase Ex. Isometric View; Ref. 8

A swerve drivebase looks something like what is shown above. Let's take a closer look at one of these swerve modules.

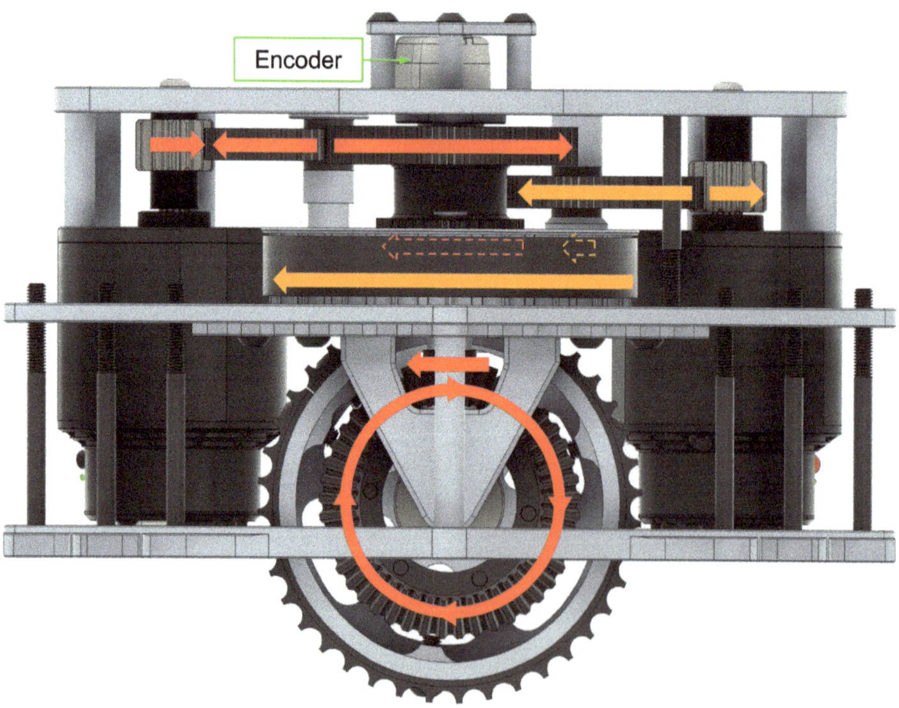


Diag. 2.13: Swerve Module Ex. Power Transmission; Ref. 8

In diagram 2.13, some structural pieces were removed to better reveal the power transmission. As depicted, a clever layering of gears and one set of pulleys enable this swerve module to function. The red set of arrows shows the full power transmission from the motor pinion to the final bevel gear that drives the wheel. The orange set of arrows shows the full power transmission from the motor pinion to a pulley that controls the orientation of the wheel. The red dotted arrow represents a gear hidden from view while the orange dotted arrow represents a pulley hidden from view. An encoder is used to measure exactly which direction the wheel is pointing when the robot starts up. From there the wheels can be calibrated to align so that all face the same direction when moving.

Since each swerve module has 1 motor for propulsion and another motor for orientation, that means 2 motors are used in each swerve module. As most swerve drives utilize a swerve module in each corner of a square drivebase, a swerve drivebase requires 8 motors in total. This could potentially be a drawback as the swerve forces teams to design the rest of the robot's mechanisms with fewer motors, but the extra maneuverability and reliability is usually considered to outweigh this drawback

Manually designing and fabricating a swerve module would be a daunting task for most FRC teams. Fortunately, there are various swerve modules available for purchase. This specific design is a kit called the MK4i from Swerve Drive Specialties.

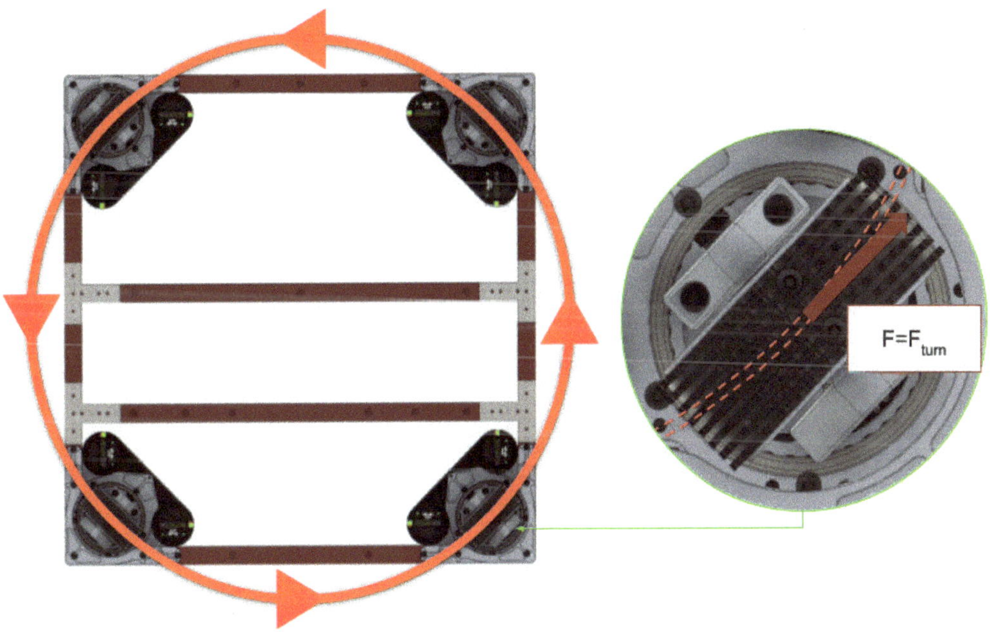


Diag. 2.14: Swerve Drivebase Ex. Turning Circle; Ref. 8

When turning, all wheels are angled at 45°. Given the wheel is able to turn so that the direction it is heading is exactly tangential to the drivebase's turning circle, all of the force applied to the ground goes into turning. This allows for the Swerve Drivebase to turn very quickly and smoothly.

## 2.9 Drivebase Gearbox Shifting

It is possible for a drivebase to be designed with dog gears to achieve gear shifting. These work similarly to the transmission system in cars. Dog gears allow the drivebase to start at a low gear, then shift to a high gear to achieve greater speed. Gearbox shifting could also be used to put the drivebase in an even lower gear than would normally be used to potentially win a pushing match, similar to how a car goes into low gear to get up a steep hill.

Theoretically, a shifting swerve gearbox is possible, though no one has published a design on this yet. Therefore, the example included in this textbook is from a WCD.

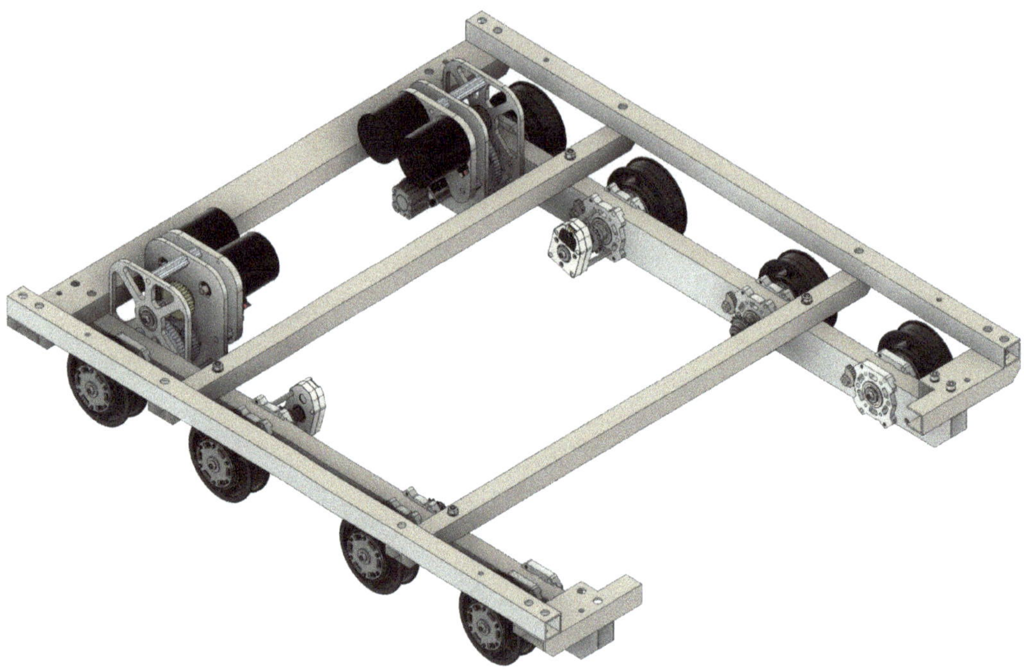

Diag. 2.15: Dog Shifting WCD Ex. Isometric View; Ref. 9

Let's zoom in to the gearbox to understand how shifting works.

As shown in diagram 2.16, a pneumatic shifts a coupler that engages either the right or left gear depending on whether the pneumatic is extended or retracted. Both gears that act as the final driven gear in the gearbox have a bearing inside them that allows them to freely spin around the gearboxes' output axle. It is only when the coupler comes into contact with one of the gears that that gear is able to drive the output shaft.

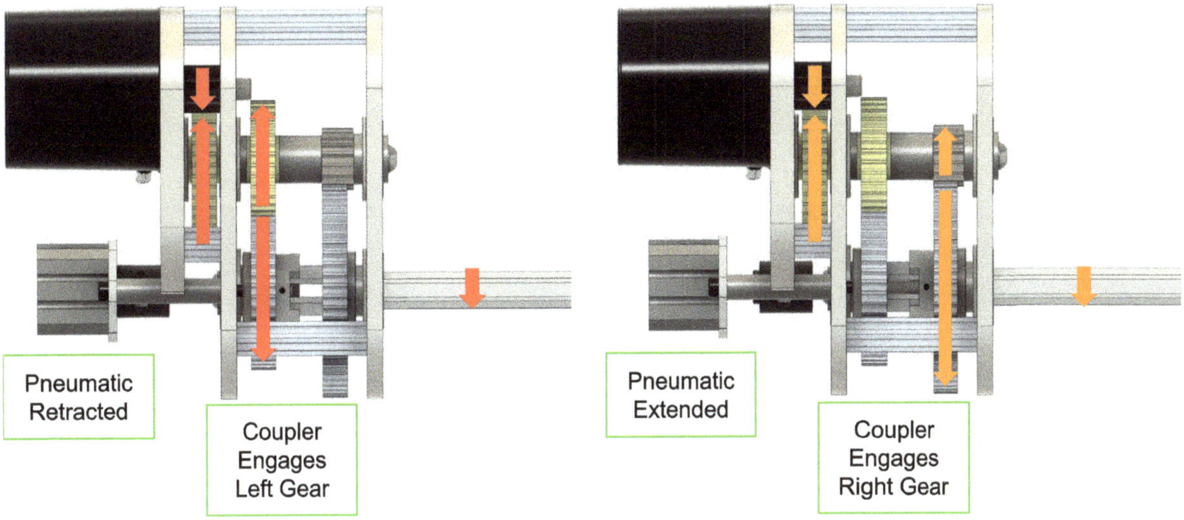


Diag. 2.16: Dog Shifting Gearbox Ex. Side View; Ref. 9

Such a design can only be achieved with unique parts, such as the dog shifting gears that have bearings in them and grooves to mesh with the coupler. For reference this example used the PTO Add On Kit from WestCoast Products.

## 2.10 Further Comparison of Drivebases

With the history of FRC drivebases covered, which drivebases are still commonly used in today's FRC games?

Since the 2024 season, FRC has made a new rule allowing only 4 propulsion motors in a drivebase. This means only 4 motors can be engaged in pushing a robot around the field. While the tank drivebase, west coast drivebase, and H-drivebase can potentially use more than 4 motors to increase speed and/or torque for more effective at defense in head-on collisions, such designs are no longer legal. In contrast, the swerve drivebase, with 4 propulsion motors and 4 orientation motors, satisfies the new rule.

Given the foregoing, almost all competitive FRC teams now use a swerve drivebase given its high maneuverability and strong defense in any direction. Complexity is minimized as FRC teams simply buy swerve module kits instead of designing and manufacturing their own swerve designs. The downsides to using a swerve are more complicated programming and a higher price tag. The other drivebase with somewhat common usage is a WCD, as this is an effective solid drivebase for rookie teams. Similar to swerve module kits, whole WCD can be bought from suppliers. The Kit of Parts (KoP) that many rookie teams use often includes a WCD.

There is in theory a niche for octocanum drivebases. If a team's design only allows for 4 motors to be allocated to the drivebase and the team wants to maintain high maneuverability, (as opposed to using simply a WCD), an octocanum drivebase would be the best option. This drivebase is complex to design and build, but is an alternative to a swerve drivebase that uses fewer motors.

## 2.11 Summary on Drivebases

The pros and the cons of the drivebases discussed in this chapter are summarised in the table on the right.

| Simple Drivebases | |
|---|---|
| • Very Simple Design<br>• Limited Maneuverability<br>• Low Cost | |
| Tank Drivebase | • Bad at turning |
| Omni Drivebase | • Good at turning<br>• Weaker against defence bots |
| West Coast Drivebase (WCD) | • Slightly more complicated design than tank<br>• Good at turning |
| **Basic Multidirectional Drivebase** | |
| • Simple Design<br>• Weak against defence bots<br>• Very maneuverable<br>• Low Cost | |
| H Drivebase | • Good at turning |
| Mecanum Drivebase | • Requires 4 independent axles<br>• Great at turning |
| **Complicated Multidirectional Drivebases** | |
| • Complex Design<br>• Very maneuverable<br>• Higher Cost Given More Parts | |
| Octocanum Drivebase | • Requires 4 independent axles<br>• Requires pneumatics<br>• Great at turning |
| Swerve Drivebase | • The complex design becomes simple if the swerve modules are bought<br>• Requires 4 independent swerve modules<br>• Requires powerful motors<br>• Complicated to program for full maneuverability<br>• Great at turning<br>• Uses the most motors (8 total) |

Diag. 2.17: Drivebase Summary Table

The drivebase is needed in all competition bots. Most teams these days choose between the WCD and the swerve drivebase, with a preference for the latter if budget and team capability allows. The flowchart in diagram 2.18 details the decision tree on the choice of drivebases.

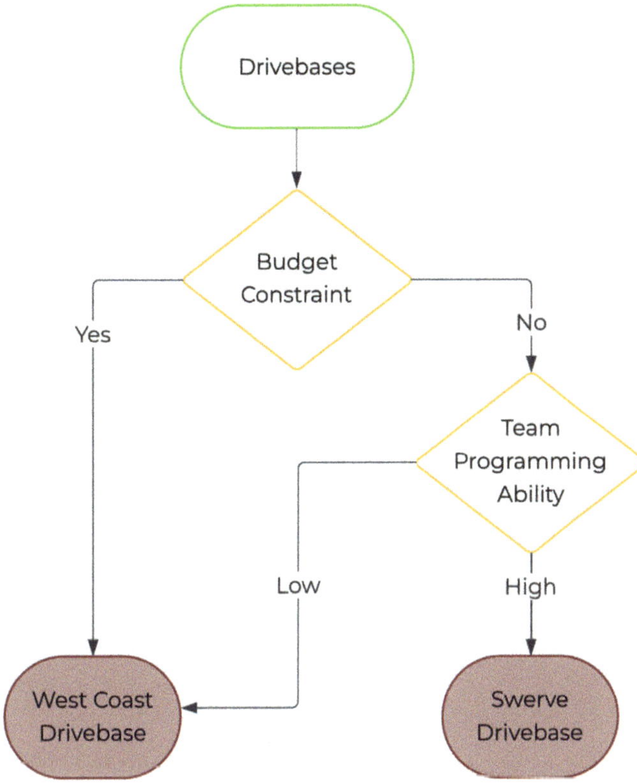


Diag. 2.18: Drivebase Flowchart

However, there may be situations where game restrictions could push teams to reconsider all the other options outlined in this chapter.

Elements of these drivebases can be incorporated into other mechanisms, like a mecanum wheel to help filter game elements (see Subchapter 3.5 on Filtering Considerations for Intakes).

The choice of drivebase does not typically impact other parts of the robot design.

# 3. Intakes

## 3.1 Introduction

Intakes are mechanisms that bring the game object into the robot. Intakes are usually part of a chain of mechanisms that a game object will move through; from the intake, through an indexer, and out via the launcher (either a flywheel or a catapult). Given this, the main job of the intake is to smoothly deliver the game object to the indexer, where the game object may be stored for some time.

These game objects tend to be balls of some size with some level of compressibility. However, intake, indexer, and launcher robots have been made to shoot discs, donuts, and frisbees for FRC games. Generally, if the game object needs to be launched by a robot, it will first enter the robot via an intake.

Unlike drivebases designs that are pretty standardized and can be recycled from season to season, intake designs change every season as FRC usually does not reuse game objects. There are many variables with how an intake should be designed based on what the game object is and how it is used in the FRC game. While this chapter provides examples on intakes and covers general ways that an intake can work, further customization beyond this textbook would be required to make an effective intake for a new FRC game.

Rollers are a key part of any intake, as they are responsible for pulling game pieces into the robot. Hard game objects require soft rollers. Soft game objects usually work better with hard rollers, but can also work with soft rollers. If a design calls for multiple rollers in a straight line, the rollers can usually be replaced by pulleys and belts.

## 3.2 Over Bumper Intakes

All FRC robots have a bumper that goes around the edge of the drivebase. While this bumper is useful for protecting the robot during collisions, it can also serve as a useful soft backboard for an intake scooping a game object in (this would be particularly good for a hard game object). On the other hand, a bumper could be considered "in the way", so an intake is designed to wrap around it. Either way, over bumper intakes are foldable intakes designed to start inside the starting legal boundaries of an FRC robot (usually this is within the frame perimeter), and then extend outside of the robot to become a fully operational intake.

Usually, it is optimal to design the over bumper intake so that it can be easily retracted. This is so that the intake can be protected during a collision with an opposing robot. If your intake gets destroyed during one bad match in an FRC competition, your team is probably limited to driving and playing defense for the rest of the competition. There are many ways to make an over bumper intake easily retractable (and extendable). The simplest is to use a pivot operated by a motor or a pneumatic.

Diag. 3.1: Over Bumper Intake Ex. 1 Isometric View; Ref. 10

This is the first example. See https://www.frc-handbook.com/ref/10 for more information on the game and full robot.

Diag. 3.2: Over Bumper Intake Ex. 1 Positions; Ref. 10

A pneumatic drives an uneven 4-bar arm (see Subchapter 7.3 on 4-Bar Arms) allowing the intake to switch between the starting and intaking positions. The outline of the pneumatic in the extended position is drawn with dotted redlines which would otherwise be hidden behind one of the plates making up the uneven 4-bar arm.

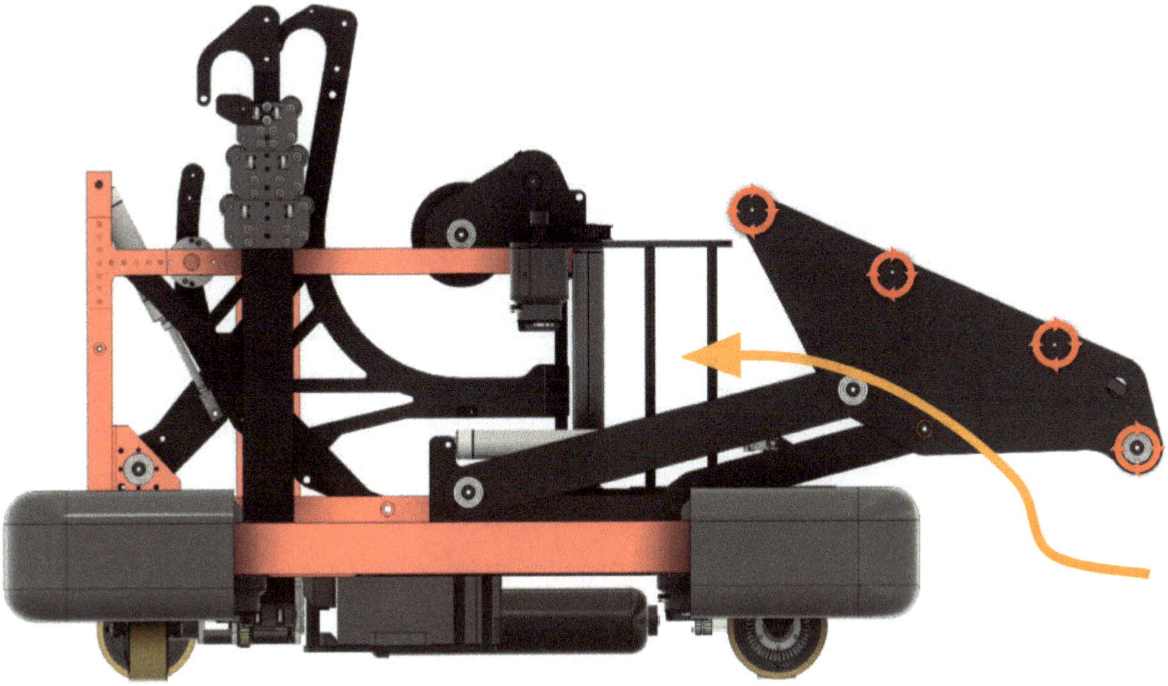

Diag. 3.3: Over Bumper Intake Ex. 1 Ball Path; Ref. 10

In the extended position, this overbumper intake is able to intake the oversized tennis balls by using 4 rollers. The red circular arrows show the direction the 4 rollers spin to intake balls. The orange arrow shows the ball path as it travels through the intake.

While these 4 rollers are all that contact the ball, a series of belts link the 4 rollers together so that they can all be driven by a single motor. The bottom right roller uses a line of small compression wheels, (instead of a traditional plastic roller), to increase the grip of the intake when first making contact with the oversized tennis balls. As can be seen, this intake also uses the bumper as a sort of backboard when intaking balls.

There are other ways to use a series of motors or series of pneumatics to make an intake retractable and extendable. Here are some examples of more complicated over bumper intake examples from Team 4817.

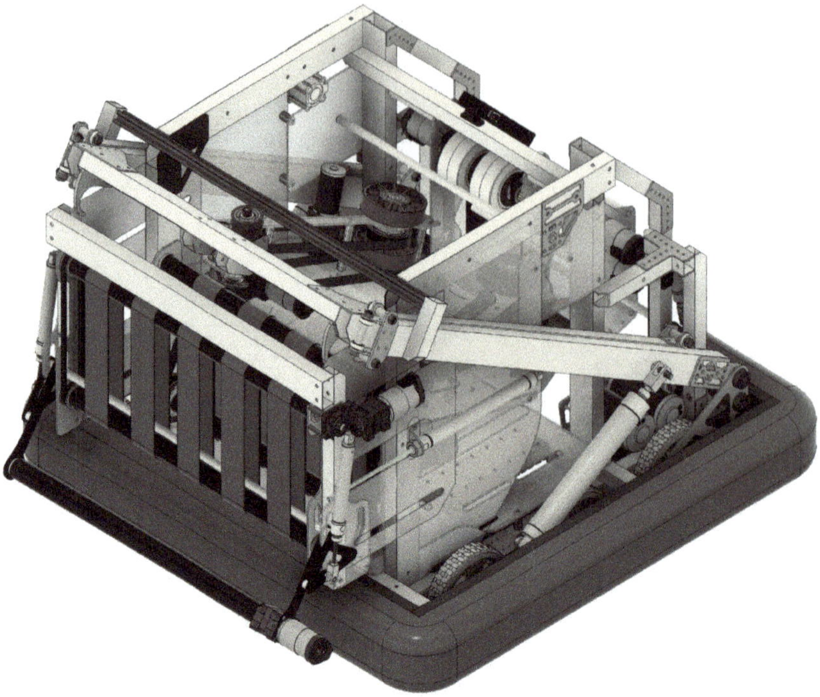

Diag. 3.4: Over Bumper Intake Ex. 2 Isometric View; Ref. 11

This is the second example. See https://www.frc-handbook.com/ref/11 for more information on the game and full robot.

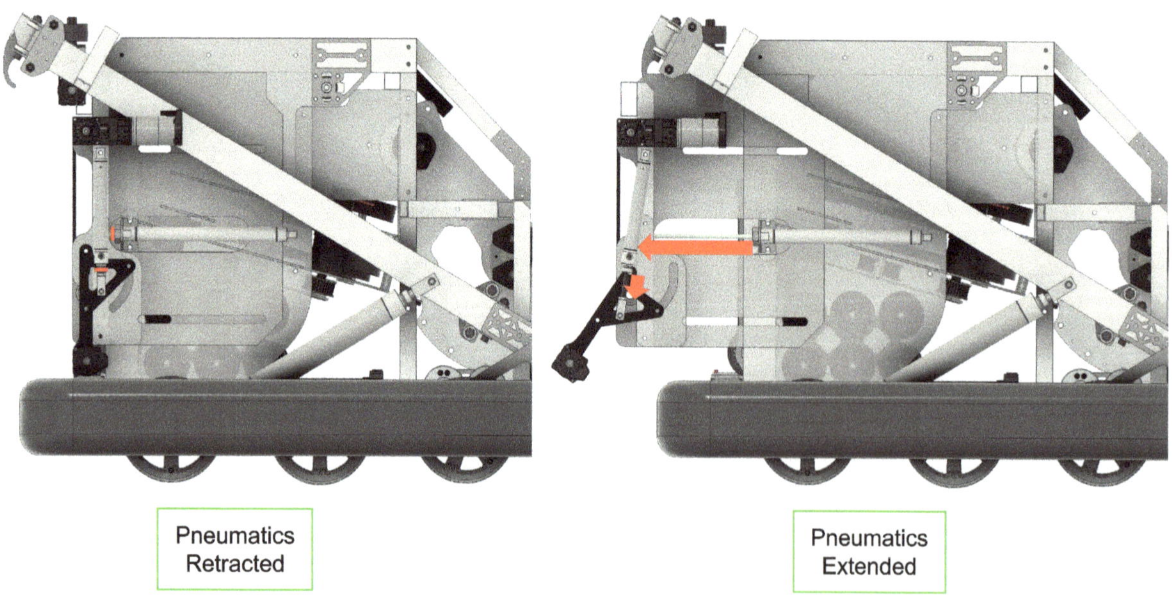


Diag. 3.5: Over Bumper Intake Ex. 2 Positions; Ref. 11

Pneumatics are used in this over bumper intake to pop out the intake when it needs to intake balls. The longer pneumatic shoves out the intake, while the smaller pneumatic gets a roller into the ideal position to intake balls.

Diag. 3.6: Over Bumper Intake Ex. 2 Ball Path; Ref. 11

With the over bumper intake extended, the ball path follows the depicted orange arrow, dropping the ball into the robot's storage/indexer. The red arrows show the movement of 2 sets of belts that pull the ball up as well as the bottom left roller's rotation to intake the ball.

This intake was designed with the intention of bringing balls high up into the robot before dropping the balls into a storage/indexer system. Intakes like this are usually used when the robot is allowed to carry many game objects. For the 2020 FRC game, the robot could carry up to 5 balls.

As noted earlier, it is possible to intentionally use the robot's bumper as part of the intake system. In this example, the bumper is used as a soft back board for the bottom left roller that first guides balls into the robot.

This intake uses 3 motors. One powers the bottom left roller, another powers the left set of belts, and a third powers the right set of belts.

Diag. 3.7: Over Bumper Intake Ex. 3 Isometric View; Ref. 8

Diagram 3.7 shows the third example. Something to note is that the intake doubles up as the storage system for this robot, as the game allowed for only one orange ring to be carried by the robot at any given time (so a larger storage is not necessary). See https://www.frc-handbook.com/ref/8 for more information on the game and full robot.

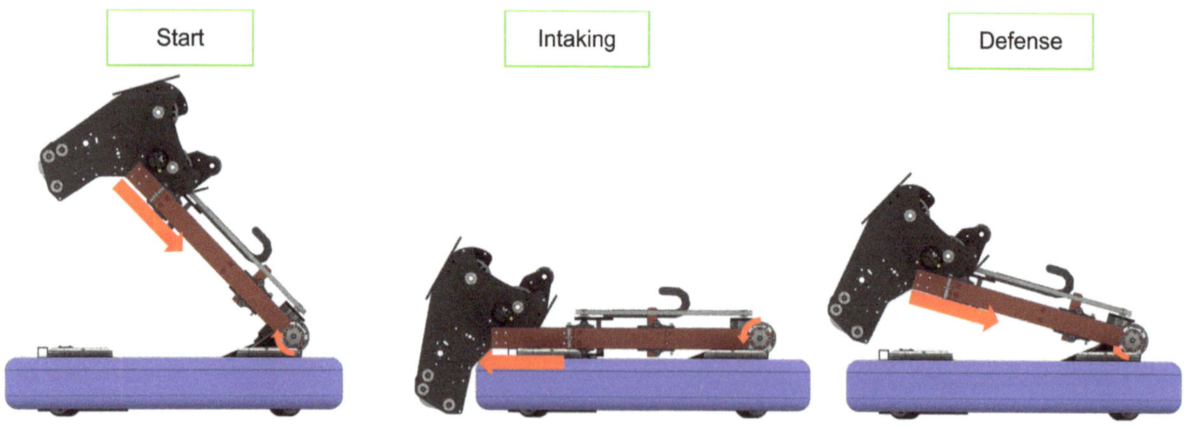


Diag 3.8: Over Bumper Intake Ex. 3 Positions; Ref 8

This robot's design had 3 key positions related to intaking. The start position shows how this robot starts every match to stay within the frame perimeter (as was required by the game rules). The intaking and defense positions show the 2 positions the robot used for the rest of the match. While the defense position is not exactly inside the frame perimeter, it was deep enough in the robot that it is hard for opposing robots to damage the intake during collisions, (notably this did happen one time during a particularly unlucky crash at regionals, which led to a frenzied repair). In this case a low center of mass for general movement is prioritised over the risk of damage in this defense position for matches.

The robot can shift between these 3 different positions by using an extendable arm (see Subchapter 9.3 for more information on this extendable arm).

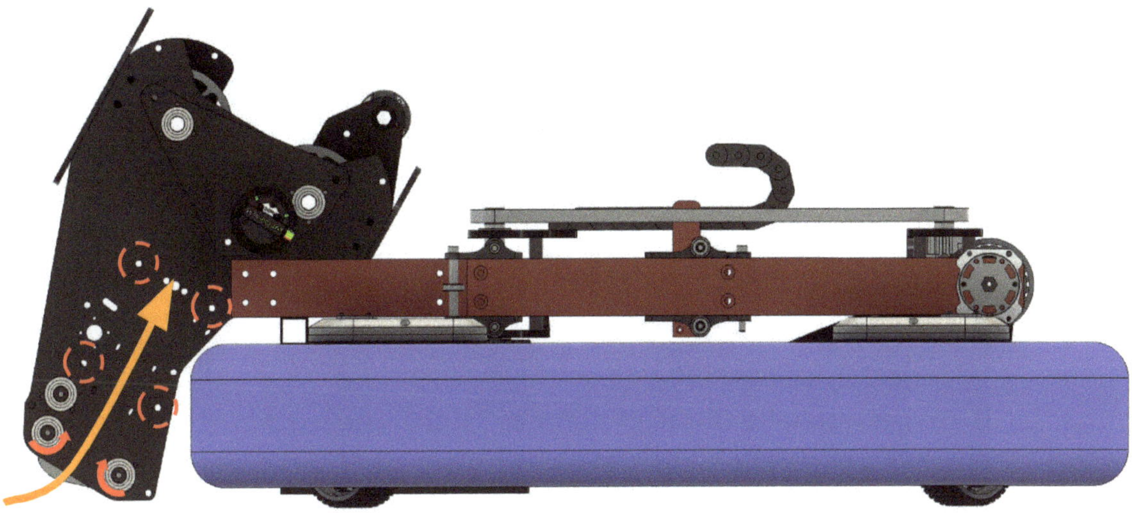

Diag. 3.9: Over Bumper Intake Ex. 3 Ring Path; Ref. 8

The ring's path for this intake is pretty intuitive. The red arrows show the rollers that spin to bring the ring in, while the orange arrow shows the ring's path. The dotted red circles are structural elements hidden from view that help guide the ring's path. The intake simply holds the ring midway through the path and only passes on the ring to the flywheel when the robot is ready to shoot the ring.

A single motor powers the entire intake. This is accomplished by a group of gears, belts, and pulleys.

## 3.3 Through Bumper Intakes

A through bumper intake works by having a gap in the bumper that the game object can pass through. Make sure when designing this type of intake that the gap in the bumper doesn't break any bumper rules for the FRC game being played.

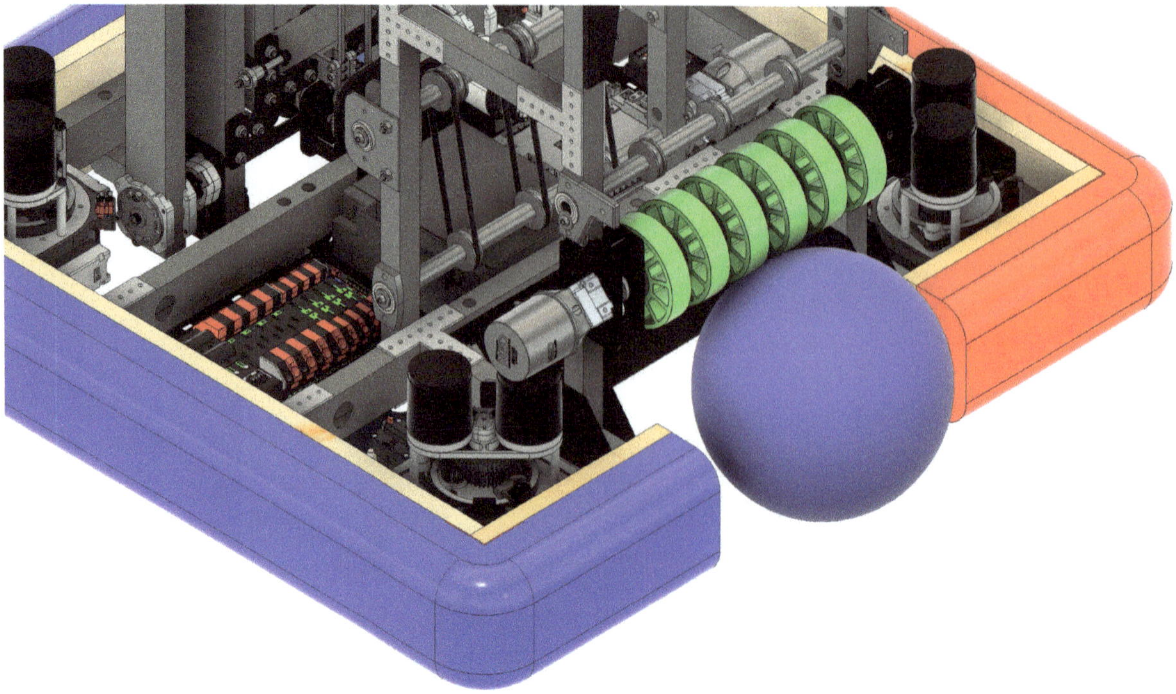

Diag. 3.10: Through Bumper Intake Ex. 1 Isometric View; Ref. 12

This is an example of a through bumber intake. See https://www.frc-handbook.com/ref/11 for more information on the game and full robot.

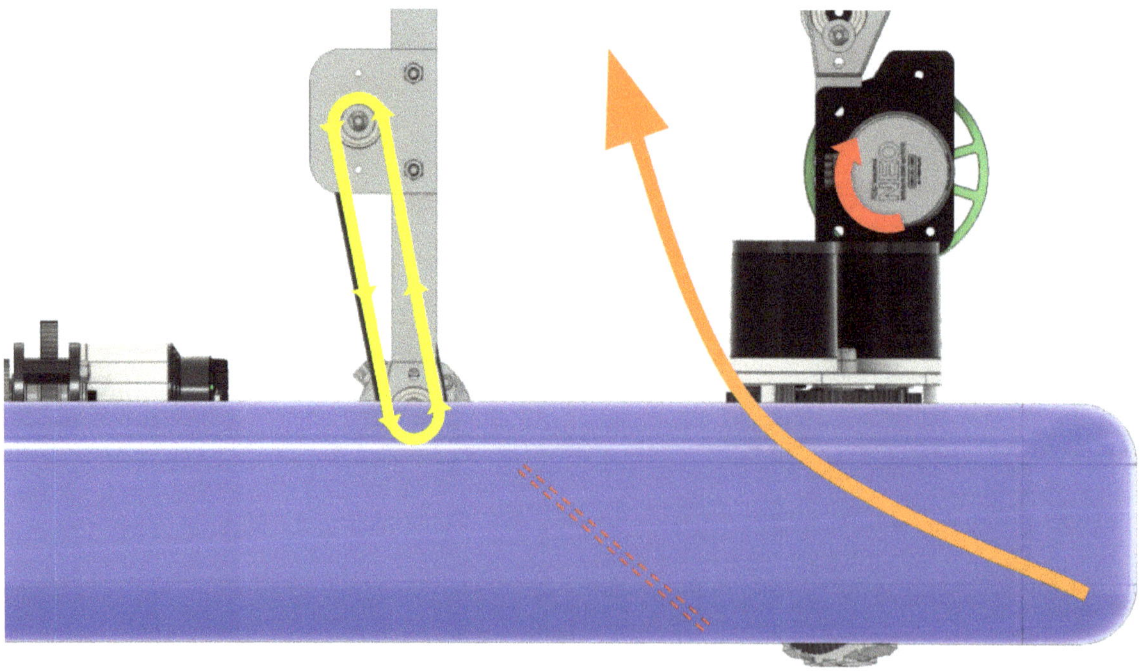

Diag. 3.11: Through Bumper Intake Ex. 1 Ball Path; Ref. 12

The ball's path in the intake follows the orange arrow. The red arrow shows the roller that engages with the ball, while the dotted red lines show relevant hidden structure that guides the ball movement. The yellow arrows show the movement of a belt that guides the ball upward, however, that section is considered to be part of the indexer of this robot.

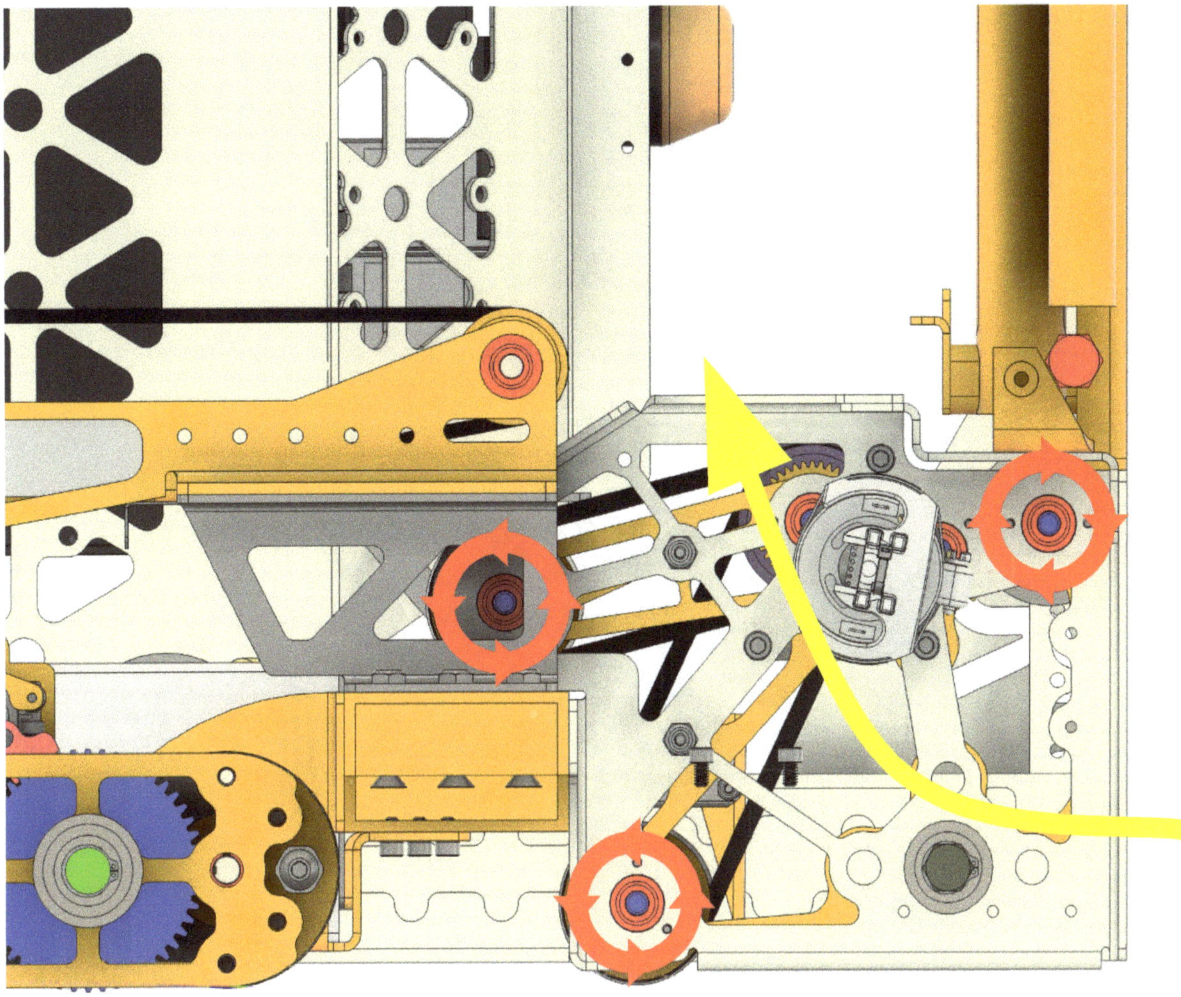

Diag. 3.12: Through Bumper Intake Ex. 2 Ball Path; Ref. 13

Refer to https://www.frc-handbook.com/ref/13 for an example of a through bumper intake from Team 118. Despite the complexity of other parts of their robot, their through bumper intake is relatively simple. It utilizes 3 rollers to bring the yellow gopher balls into the robot. The direction of the rollers is shown by the red circular arrows, while the yellow arrow shows the ball's path through the intake.

This intake is powered by a single motor and uses a group of belts and gears to turn all 3 rollers.

Through bumper intakes are the most simple intake design since extension and retraction mechanisms need not be factored into the design, (unlike overbumper intakes). Due to the protected location of this intake, it is nearly impossible for opposing robots to damage it during collisions. However, a through bumper intake usually means a narrower intake as the intake has to comply with bumper rules. This means the driver has to more accurately align the robot with the game object for through bumper intakes. (See Subchapter 3.5 on Filtering Considerations for Intakes.)

## 3.4 Under Bumper Intakes

An under bumper intake works by passing the game object under the robot's bumper before feeding it into the robot. In some FRC games such an intake would not be possible given the size of the game object. FRC bumpers have rules about how high off the ground they can be, limiting the clearance between the ground and the bumper.

Here is a notable example taken from the robot reveal video by University of Minnesota Robotics that utilizes an under bumper intake.

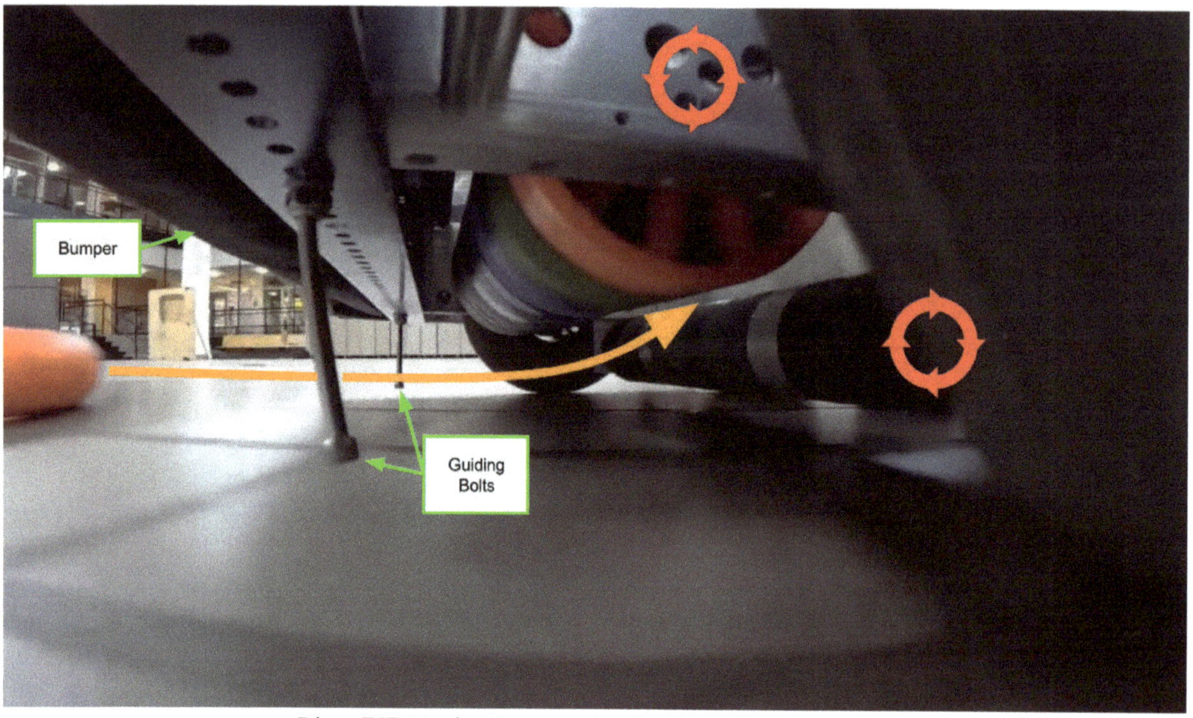


Diag. 3.13: Under Bumper Intake Ex. Ball Path; Ref. 14

Refer to https://www.frc-handbook.com/ref/14 for this Robot in 3 Days (RI3D) video example. The robot uses a recessed intake somewhat similar in design to Reference 12's intake. 2 bolts are used to guide the ring properly into the intake. Rollers shown by the red circular arrows guide the ring along the orange ring path.

The robot can only intake by driving over rings in one direction. It is theoretically possible to design an intake that could intake the rings under the bumper from multiple directions, though this is a lot harder to do. A more fully developed under bumper intake example can be seen in the 2024 robot reveal video by Team 4613 at https://www.frc-handbook.com/ref/15.

An under bumper intake has almost all of the same general benefits and drawbacks as a through bumper intake. There are a few slight differences. For some FRC games, a through bumper intake is not possible due to bumper rules requiring the bumper to wrap fully around the robot's frame perimeter. Even if a complete bumper is not required for an FRC game, utilizing an under bumper intake may still be advantageous since a complete bumper offers better protection during collisions with opposing robots. However, keep in mind that an under bumper intake is only possible if the game object is short enough to pass under a legal bumper's mounting height.

## 3.5 Filtering Considerations for Intakes

Intakes are almost always part of a chain of mechanisms: intake to indexer to launcher. In general, a robot is considered to be better designed if it is easier to drive. One way to make a robot easier to drive is to have a very wide intake. However, this creates a potential problem for the robot's launcher. Before the game object reaches the launcher it is critical to align the game object to make shots by the launcher more predictable and accurate. Considering this, somewhere along this chain of mechanisms, usually a filter of some sort is required to accomplish this alignment action.

It is possible for the filter to be in the indexer (see Subchapter 4.5 on Filtering Considerations for Indexers). This subchapter provides examples of filtering done at the intake mechanism.

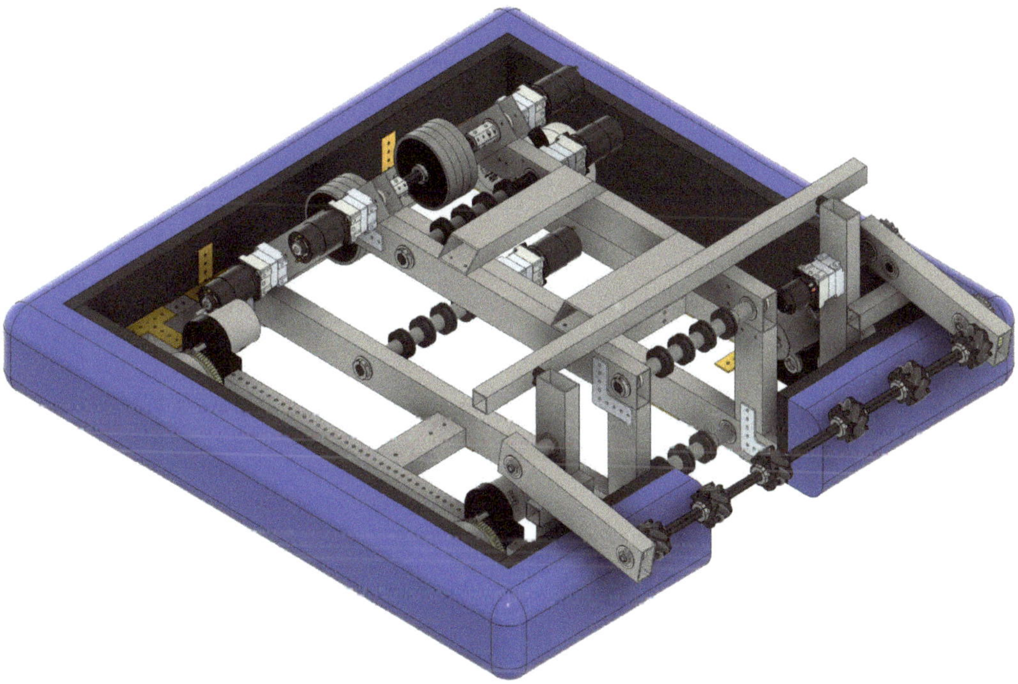

Diag. 3.14: Over Bumper Intake Ex. 4 Isometric View; Ref. 1

For Reference 1, see https://www.frc-handbook.com/ref/1 for more information on the game and full robot. A feature of this robot is that it has a very wide intake that theoretically filters balls towards the center. This is done by using mecanum wheels. While the wheels will push the balls forward, the mecanum wheels will also push the balls sideways. Using the correct type of mecanum wheel will direct the balls to move toward the center.

However, some theoretical things do not always work as anticipated. Post CAD of this design, an additional structure had to be added to stop the ball from being pulled over the bumper by the roller. It was easier for the ball to move over the bumper than strafe sideways—until that path was blocked off by the addition of a wall.

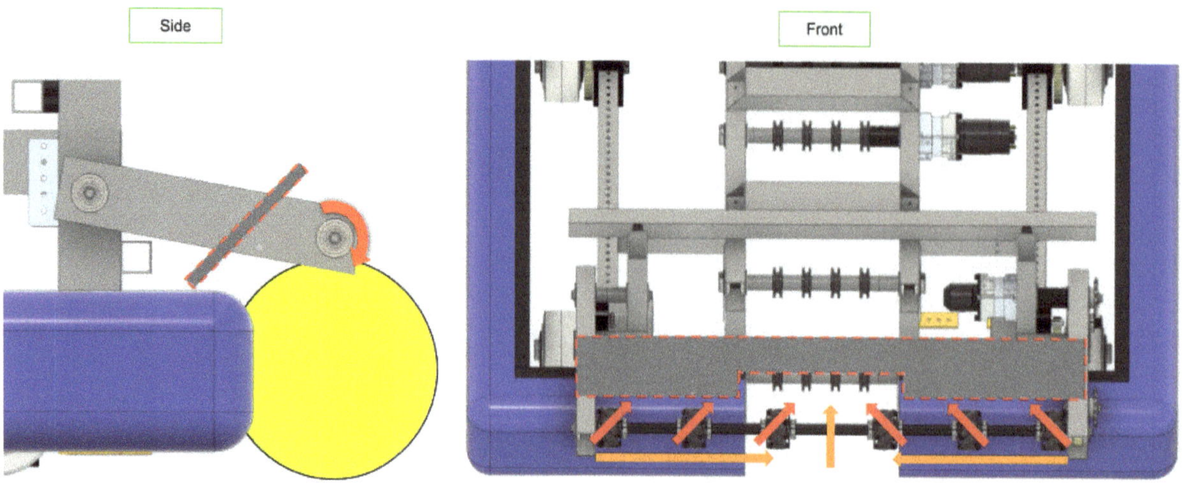


Diag. 3.15: Over Bumper Intake Ex. 4 Ball Path; Ref. 1

The yellow circle represents the yellow balls used for the 2021 FRC game. The grey structure outlined by a dotted red line shows the structure that was added post CAD. The red arrow in the side view shows the direction that the roller rotates. The red arrows in the front view show the direction towards which the mecanum wheels push the ball while the ball is in contact. The balls are directed to follow the paths shown by the orange arrows, thereby filtering the ball towards the center of the robot.

There are other ways to design an intake that filters game objects. Such examples include sideways rollers that push the game objects across the robot until the game object reaches the center or side. A conceptual example of such a sideways filtering intake by Capital Robotics, as featured on the FUN Robotics Network, can be found at https://www.frc-handbook.com/ref/16.

Sometimes the filtering is done outside of the intake by a physical structure.

Diag. 3.16: Through Bumper Intake Ex. 1 Filtering; Ref. 12

In Reference 12, the robot does not have a very wide intake. In effect, the bumper acts as a filter for this intake. This makes it harder for the driver to pick up balls, but it also means that any ball that has been picked up by the robot is automatically centered to be shot accurately by the robot's flywheel.

Not exactly an optimal filtering technique, but a very basic and easy one to implement.

## 3.6 Summary on Intakes

Here's a quick recap of the intakes discussed:

| Over Bumper Intake |
| --- |
| • Complex to design<br>• Works with all game objects<br>• Retractable/extendable<br>  ○ Either linear slide action or rotating pivot; uses pneumatics or motors<br>  ○ Vulnerable to collisions when extended<br>• Bumper as a soft back board or extends a backboard with the intake<br>• Allows for larger storage<br>• Filtering considerations (intake ➢ indexer ➢ launcher)<br>  ○ Usually filters at indexer<br>  ○ Can filter at intake with use of sideways rollers or mecanum wheels |
| **Through Bumper Intake** |
| • Simple to design<br>• Works with almost any game object<br>• Recessed (protected from collisions)<br>• Compact<br>• Gap in bumper often acts as a filter<br>• Intake width limited by gap between relevant drivebase parts<br>• Filtering considerations (intake ➢ indexer ➢ launcher)<br>  ○ Usually filters at intake due to limited width |
| **Under Bumper Intake** |
| • Simple to design<br>  ○ Gets complicated if drivebase is multi-directional<br>  ○ Usually harder to filter than through bumper intake<br>• Works only with small or flat game objects that can pass under a bumper<br>• Recessed (protected from collisions)<br>• Filtering considerations (intake ➢ indexer ➢ launcher)<br>  ○ Filtering at intake ensures a more effective capturing of game elements<br>  ○ Filtering at indexer is simpler |

Diag. 3.17: Intake Summary Table

If looking at intakes in isolation, the following flowchart provides a way to analyze which intake to use in a robot design. Note in reality, often all mechanisms from intakes, indexers, launchers, claws etc. are considered at the same time during robot design. In particular, decisions about the indexer have to be considered in tandem with the intake.

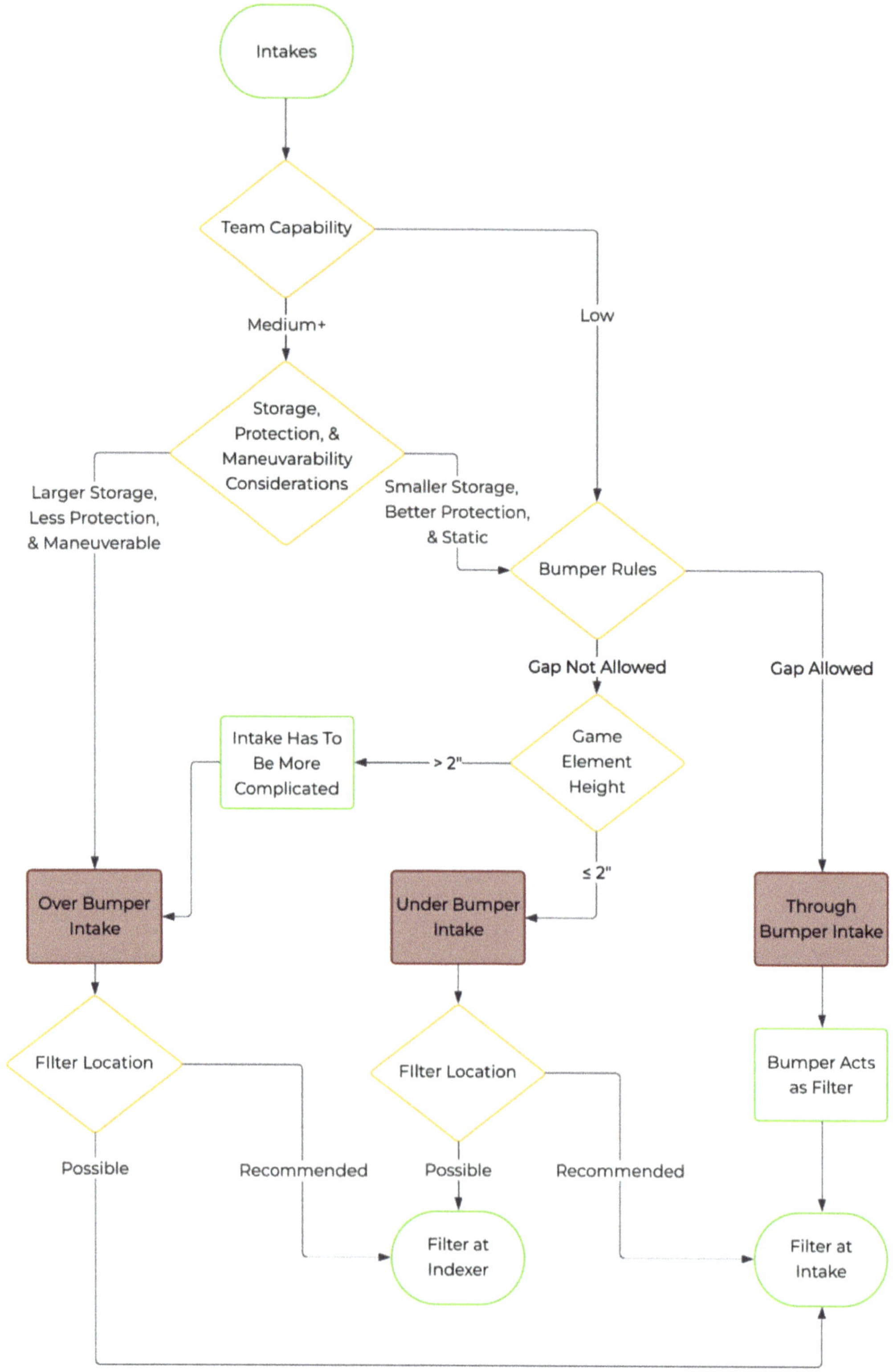


Diag. 3.18: Intake Flowchart

Note that this flowchart includes the most common factors used to specify the intake. Additional factors may apply in specific FRC games. For example, the 2024 FRC game rule disallowed gaps in bumpers, thus removing the option for a through bumper intake.

It is therefore important to have a strong understanding of the rules for each year's game as potential rules may preclude the use of a specific mechanism recommended here. A key design decision is determining where to filter the game element—at the intake or indexer. When designing launching type robots, it is important to provide a good path for the game object, including effective filtering from the intake to the indexer to the launcher.

As can be seen in the Intake Flowchart, the intake is designed in tandem with the indexer and the filtering mechanism.

# 4. Indexers and Storage Systems

## 4.1 Introduction

Indexers are mechanisms that propel game objects from the intake to the launcher (either a flywheel or a catapult). Indexers usually act as the robot's storage mechanism. Until the robot is ready to launch the game object, the indexer keeps hold of the game object. Once the robot is ready to shoot, the indexer quickly passes on the game object to the launcher.

As noted in Subchapter 3.1 there is decent variability for game objects.

There is not a set way that an indexer should be designed given this variability in game objects. This chapter provides examples of the different types of indexers that have been used in previous FRC games.

The most critical challenge in designing an indexer is preventing game objects from jamming as they travel along the indexer. This is something to pay attention to when designing the details of the indexer but will not be dealt with here.

Certain indexer designs require a "feeder" to redirect a game object from the indexer towards the flywheel. The "feeder" in the context of the indexer is an additional assembly that changes the direction of the game object so as to facilitate the smooth exit of the game object from the indexer into the flywheel.

## 4.2 Linear Indexers

A linear indexer is the most simple of Indexers. A series of rollers or belts bring a game object in a straight path from the intake to the launcher (flywheel or catapult). This may be horizontally across the robot or vertically up the robot. Typically it makes sense to use this kind of indexer when the robot is not allowed to carry many game objects and the game object gets filtered before reaching the indexer.

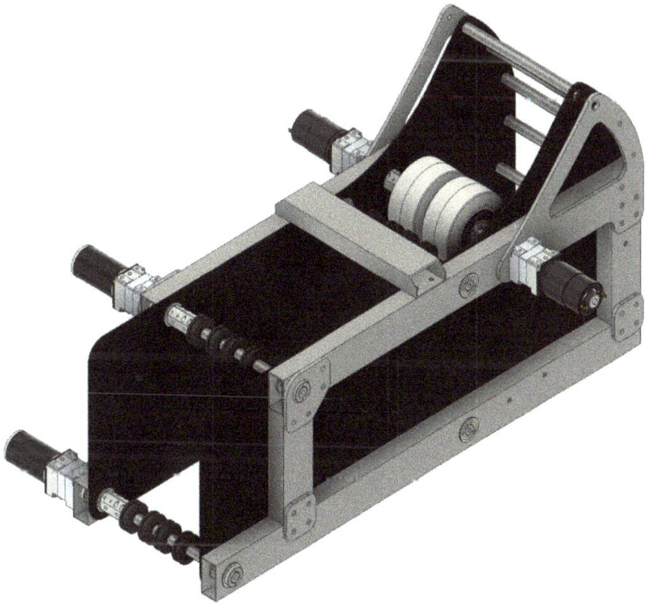

Diag. 4.1: Linear Indexer Ex. 1 Isometric View; Ref. 17

This is a horizontal linear indexer example. For more information on the game and full robot see https://www.frc-handbook.com/ref/17.

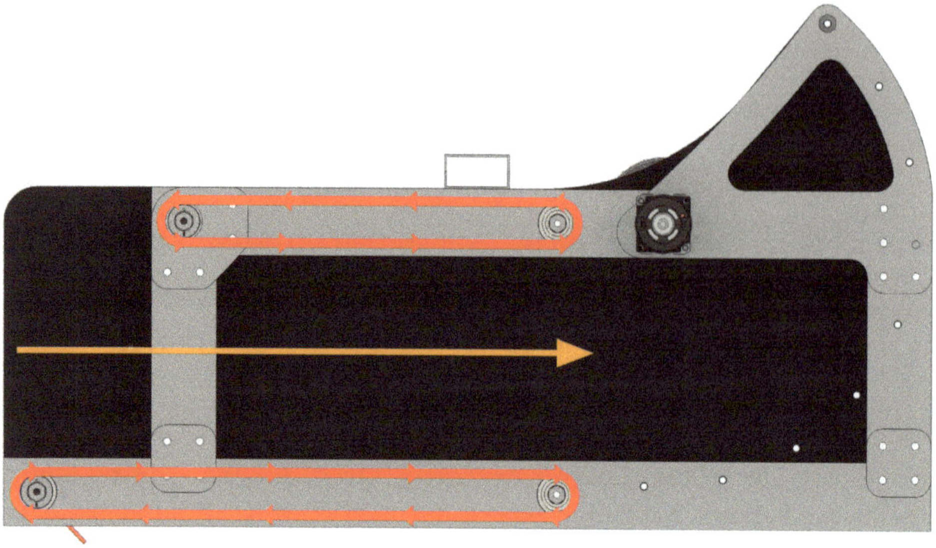

Diag. 4.2: Linear Indexer Ex. 1 Ball Path; Ref. 17

The ball's path for this indexer is linear. The red arrows show the direction that the belts spin to move the ball along, while the orange arrow shows the ball's path. These 2 sets of belts allow the indexer to bring the ball from the intake to the flywheel. One motor powers the top set of belts while another motor powers the bottom set of belts.

Diag 4.3: Linear Indexer Ex. 2 Isometric View; Ref. 12

Diagram 4.3 shows a vertical linear indexer example. For more information on the game and full robot see https://www.frc-handbook.com/ref/12.

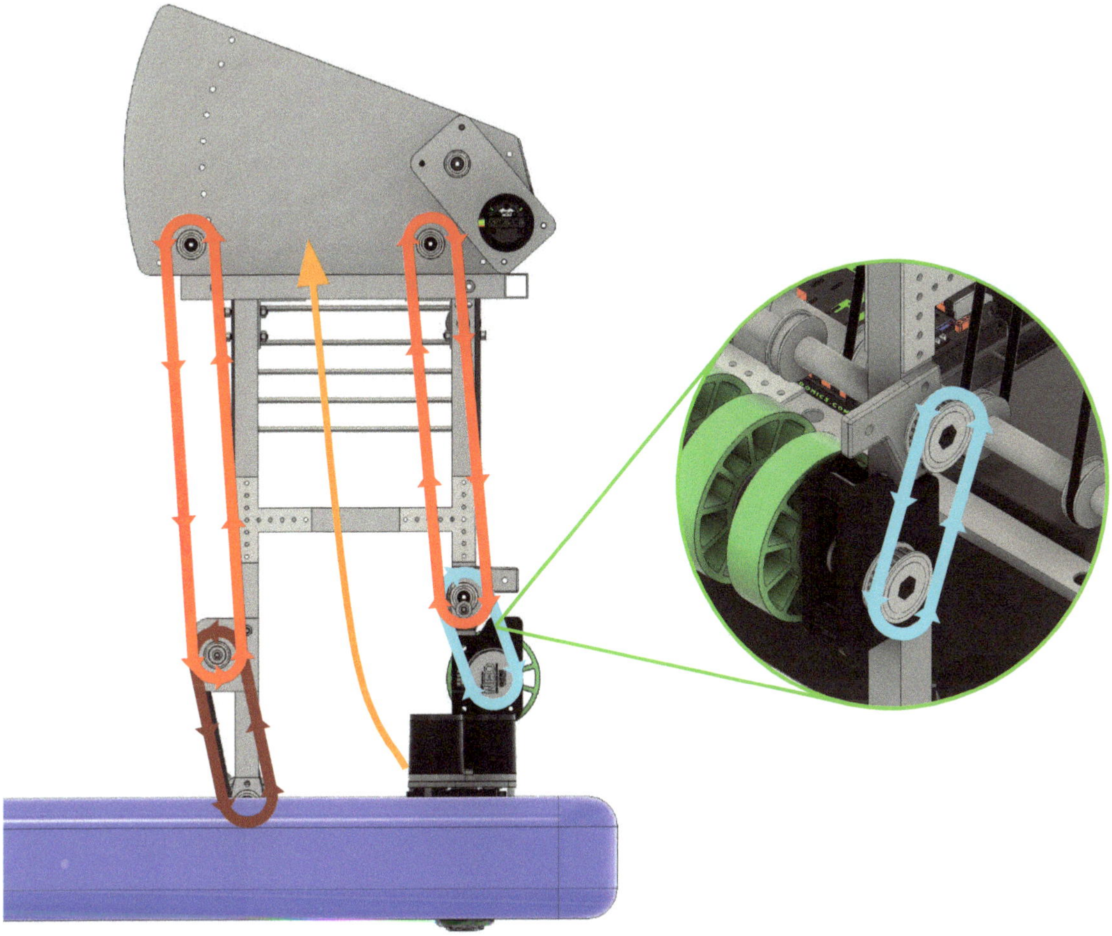

Diag. 4.4: Linear Indexer Ex. 2 Ball Path; Ref. 12

The ball's path for this indexer is very nearly linear. The red arrows show belts that spin to move the ball along, while the orange arrow shows the ball's path. This set up allows the indexer to bring the ball from the intake up to the flywheel. The darker red arrows indicate the 2nd set of belts wrapped around the same axle on the left to guide the ball upward. One motor powers the left set of belts. The motor for the intake powers the right set of belts via an additional belt represented by the cyan arrows. The green zoom in shows that this additional belt is not in the same plane as the rest of the belts and does not touch the ball.

## 4.3 Moving Floor Indexers

A moving floor indexer is best for a game where the robot is allowed to hold many game objects. It acts as a very large storage container with the moving floor pushing all of the game objects to one side of the storage area. The objective is to create a consistent line of game objects. Positioned underneath the line is a linear indexer that then moves the game objects into a launcher (usually a flywheel, but could be a catapult).

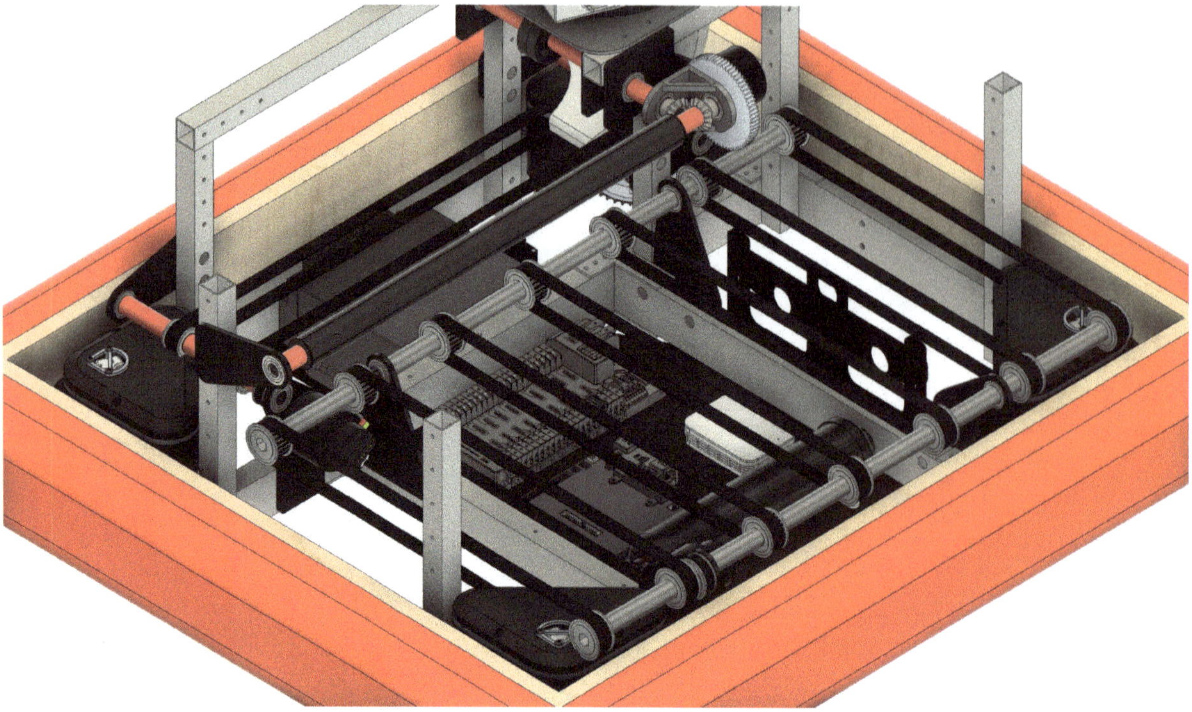

Diag. 4.5: Moving Floor Indexer Ex. 1 Isometric View; Ref. 18

This is the best example of a moving floor indexer that Team 4817 has built. See https://www.frc-handbook.com/ref/18 for more information on the game and full robot.

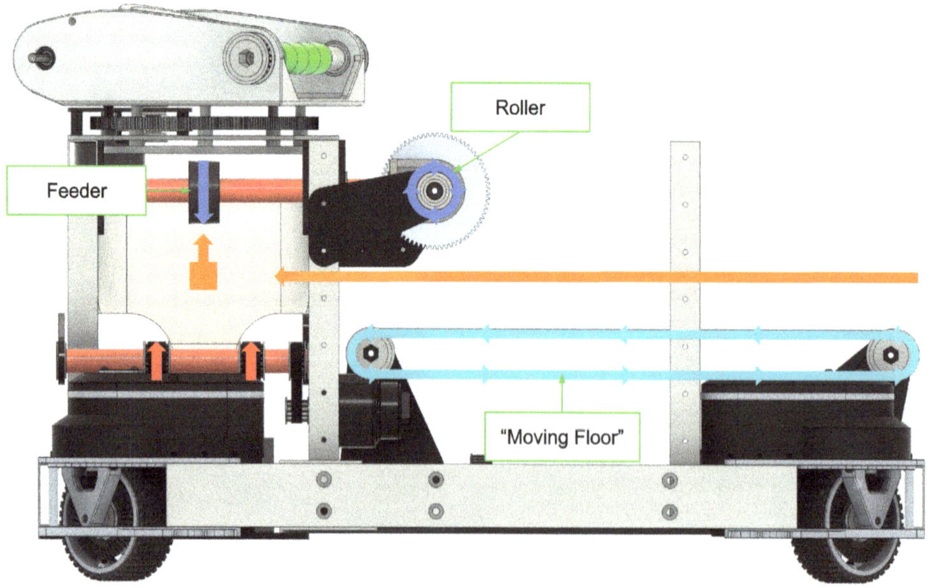


Diag 4.6: Moving Floor Indexer Ex. 1 Ball Path Left View; Ref 18

In diagram 4.6, the bottom right set of cyan arrows shows the direction in which the set of belts rotates. This set of belts makes up the "moving floor" of the indexer. The top right roller guides balls into and keeps them in the left section of the robot. This allows the balls to follow the path drawn by the long orange arrow. On the left side of the robot is a linear indexer that moves the yellow gopher balls into the flywheel. The 2 bottom red arrows show the direction the 2 belts rotate to move along the balls, with the top left blue arrow showing the direction the feeder spins to propel the balls into the flywheel.

The bottom right set of belts is powered by one motor, colored with the cyan arrow set. The top left feeder wheel and top roller is powered by a different motor, colored light blue. Both axes are powered by one motor through the use of bevel gears. The bottom left pair of belts is powered by a third motor, colored with red arrows.

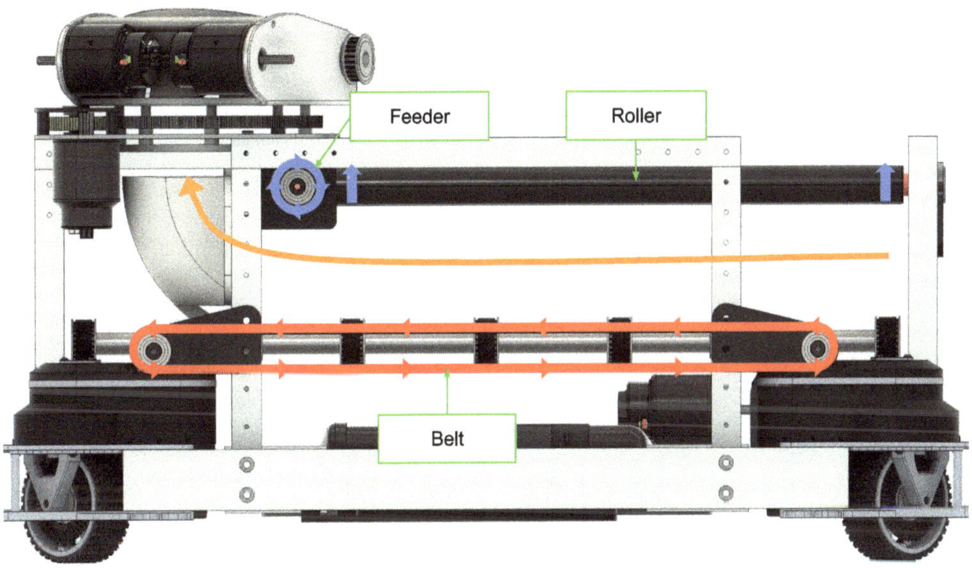


Diag. 4.7: Moving Floor Indexer Ex. 1 Ball Path Back View; Ref. 18

Diagram 4.7 shows the indexer from another view. The bottom set of belts drawn with red arrows matches the belts located in the bottom left of diagram 4.7. The top left blue circular arrows show the direction the feeder spins to bring balls up into the intake. The orange arrow shows the ball's path.

Refer to https://www.frc-handbook.com/ref/13 for another example of a moving floor indexer. There are similarities between Reference 18 and Team 118's moving floor indexer. In fact, Team 4817 took inspiration from Team 118 when designing Reference 18. The next diagram highlights a part of Team 118's design that makes their moving floor indexer superior.

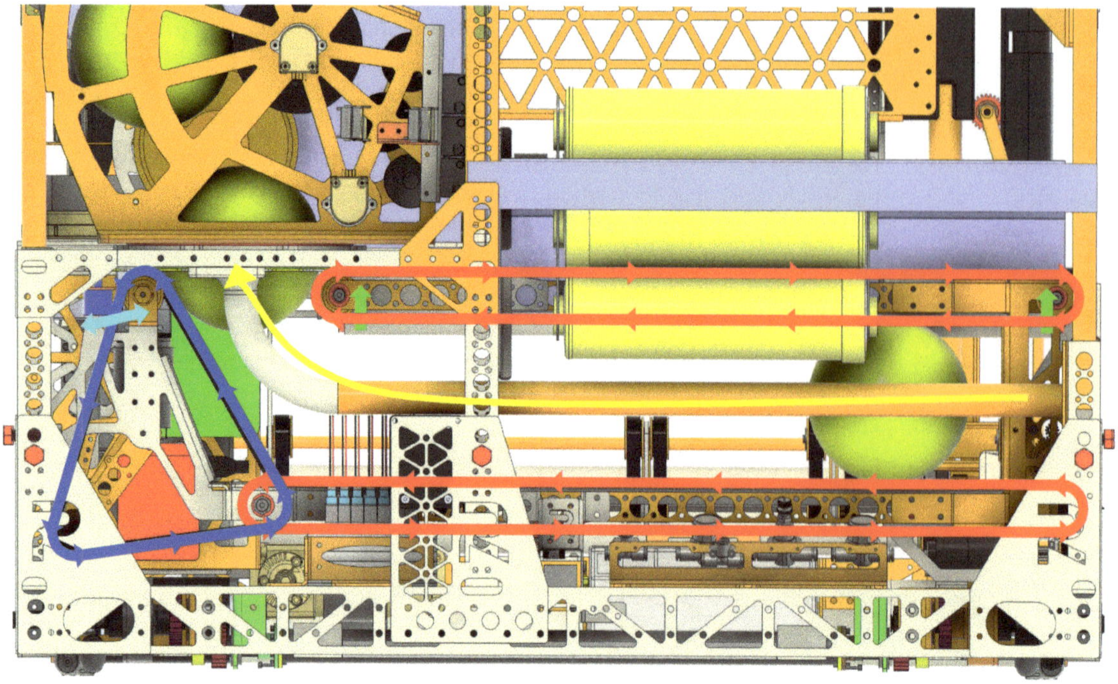

Diag. 4.8: Moving Floor Indexer Ex. 2 Ball Path Back View; Ref. 13

The first difference is that in this moving floor indexer, there are belts on both the top and bottom, as shown by the bright red set of arrows. The second and more distinctive difference lies in how this indexer's feeder works. A pair of belts wrap around a set of pulleys forming a triangle, shown by the dark blue triangular arrows. Their current position is held in place by a spring, but when a ball passes by that area, the spring gives way to open the top just wide enough for a ball to get past, as shown by the cyan arrows. This design forces the ball to stay in the flywheel once it reaches it, and allows the triangular belts to maintain better contact with the ball as it curves up into the flywheel. These 2 differences lead to a smoother and more reliable ball path than Reference 18, as shown by the yellow arrow. A similarity between the two designs is in the spinning roller that propels the balls into the linear indexer as shown by the tiny green arrows behind the top right set of belts.

## 4.4 Rotating Indexers

A rotating indexer moves game objects along a circular path. The rotation caused by the rotating indexer creates a centrifugal force that pushes balls out of the indexer and into a launcher (usually a flywheel, but could be a catapult). Sometimes, a barrier is put in place to prevent game objects from leaving the rotating indexer until the launcher is ready to shoot.

Given how a rotating indexer works, game objects usually need to be dropped on top of the rotating indexer, meaning an intake that brings game objects high up into the robot is required.

Once the game object leaves a rotating indexer, an additional wheel is used to push the game object into the flywheel.

Diag. 4.9: Rotating Indexer Ex. 1 Isometric View; Ref. 11

This is an example of a rotating indexer by Team 4817. For more information on the game and the full robot see https://www.frc-handbook.com/ref/11.

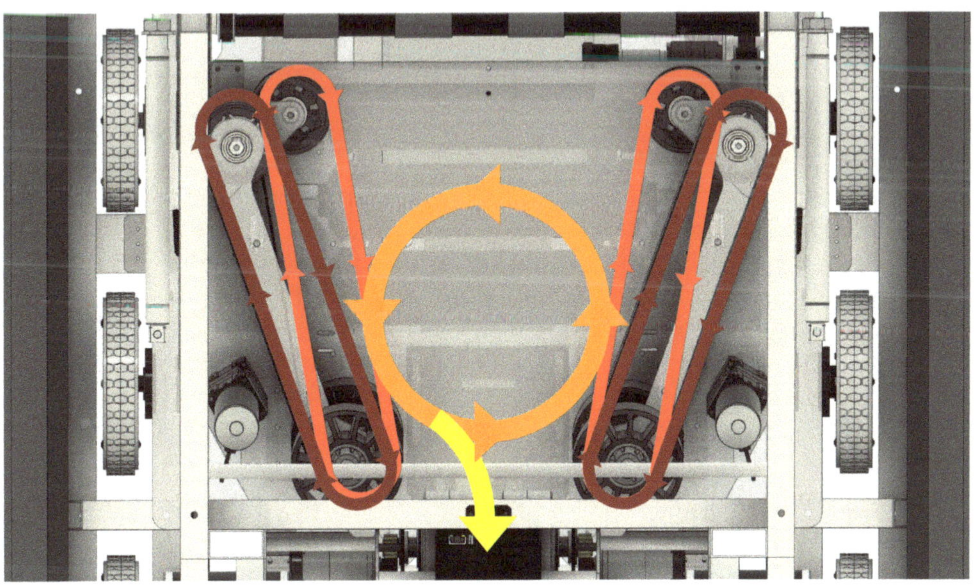

Diag. 4.10: Rotating Indexer Ex. 1 Ball Path; Ref. 11

This rotating indexer works by spinning one set of belts toward the flywheel, while the other set of belts spin away from the flywheel, as depicted by the red arrows. The plate supporting the indexer slopes downwards towards the flywheel to encourage game objects to roll in that direction. The slope, combined with the rotating indexer keeps balls in a constant spin to prevent jamming, as depicted by the orange circular arrows. When the robot is ready to shoot, the right side of the belts can be momentarily halted or spun in the reverse direction to allow a ball to pass through. This creates the feeding action that allows the balls to follow the yellow arrow into the flywheel.

A separate motor drives each side of the belts, allowing for independent control.

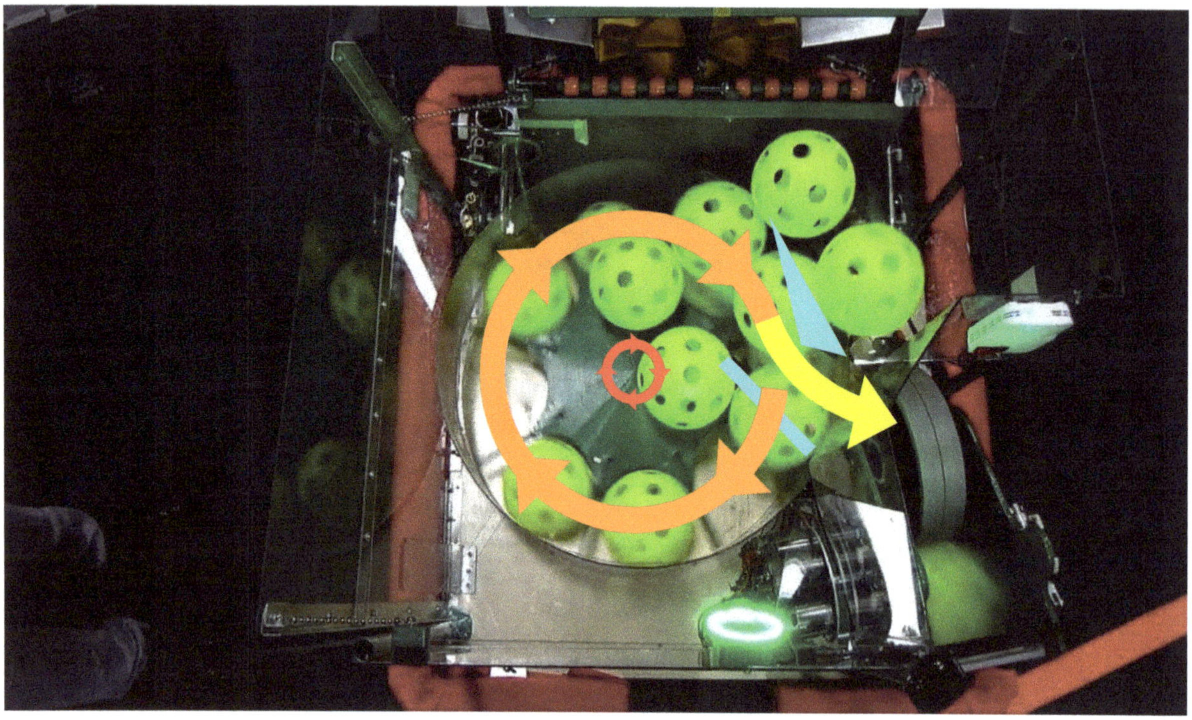

Diag. 4.11: Rotating Indexer Ex. 2 Ball Path; Ref. 19

Refer to https://www.frc-handbook.com/ref/19 for another example of a rotating indexer, designed by Team 1986. Watching Team 1986's robot reveal video will add further clarity to how this type of rotating indexer works. This turntable moves clockwise, as shown by the circular red arrows. The dark green turntable is fitted with evenly spaced flippers, each designed to fit a single gopher ball. Once a gopher ball falls into one of these flipper-shaped placeholders, it follows the path marked by the orange arrows. Eventually, the gopher ball interacts with the path guides of the feeder, as marked in cyan. The feeder then diverts the gopher ball along the yellow path to the flywheel.

The turntable rotates only when the robot is ready to fire gopher balls. Otherwise, the turntable is stationary. Due to this clever design detail, jamming is not a problem. If balls do not fall into the slits when the turntable is stationary, it will do so once the turntable starts to move again.

It's unclear how many motors are used for each part of this design, given a full CAD is not available. With the strength of modern FRC motors, one motor would be enough to power the turntable, and another motor would power the feeder under the flywheel that helps to bring the balls up to the flywheel.

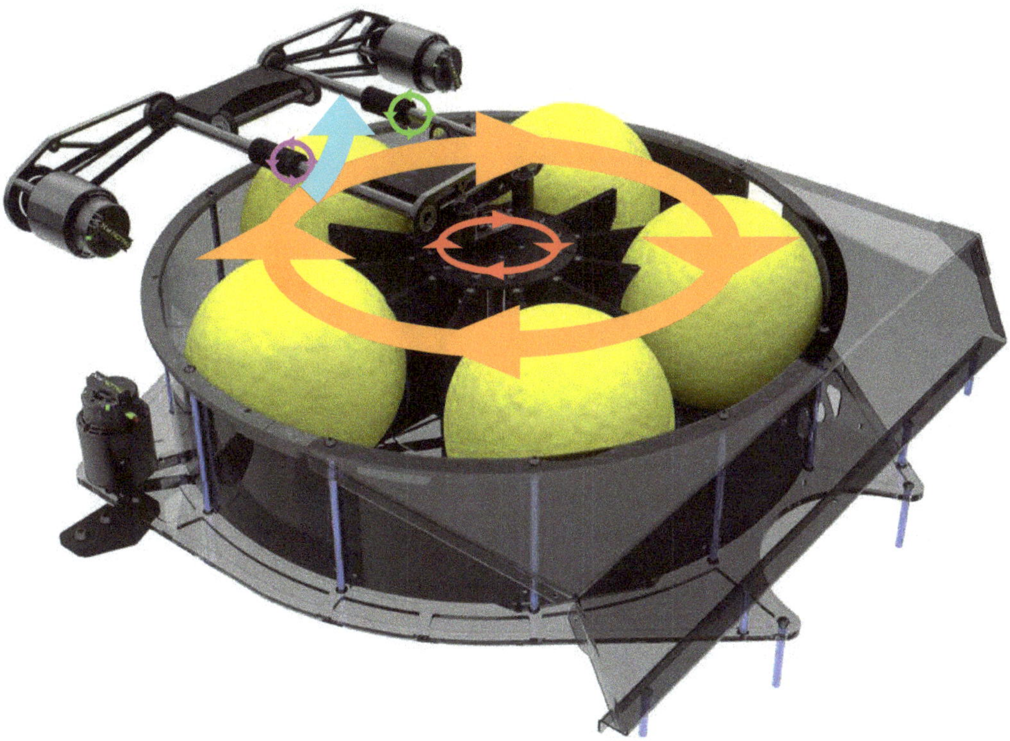

Diag. 4.12: Rotating Indexer Ex. 3 Ball Path; Ref. 20

See https://www.frc-handbook.com/ref/20 for more information on the example from Team 254. This rotating indexer is somewhat similar to Reference 19. A turntable spins, as shown by the center red circular arrows. Balls fall into an area that is at the outer edge of the turntable, where semi-flexible flippers are able to spin the balls shown by the orange circular arrows. To continue this motion, a feeder spins in the directions shown by the 2 smaller sets of arrows near the top left of diagram 4.12. When the feeder needs to pull a ball out of the rotating indexer, (and into the flywheel above it), the left side of the feeder continues to spin in the direction shown by the purple set of arrows, while the right side of the feeder spins in the reverse direction shown by the green set of circular arrows. This allows a ball to exit the rotating indexer following the large cyan arrow.

One motor powers the turntable, another motor powers the left side of the feeder, and a third motor powers the right side of the feeder.

## 4.5 Filtering Considerations for Indexers

A moving floor indexer and a rotating indexer act as effective filters before the game object reaches the flywheel. This chapter provides yet another example of filtering at the indexer.

Diag. 4.13: Linear Indexer Ex. 3 Isometric View; Ref. 10

This is an example of a filter in front of a linear indexer. For more information on the game and full robot see https://www.frc-handbook.com/ref/10.

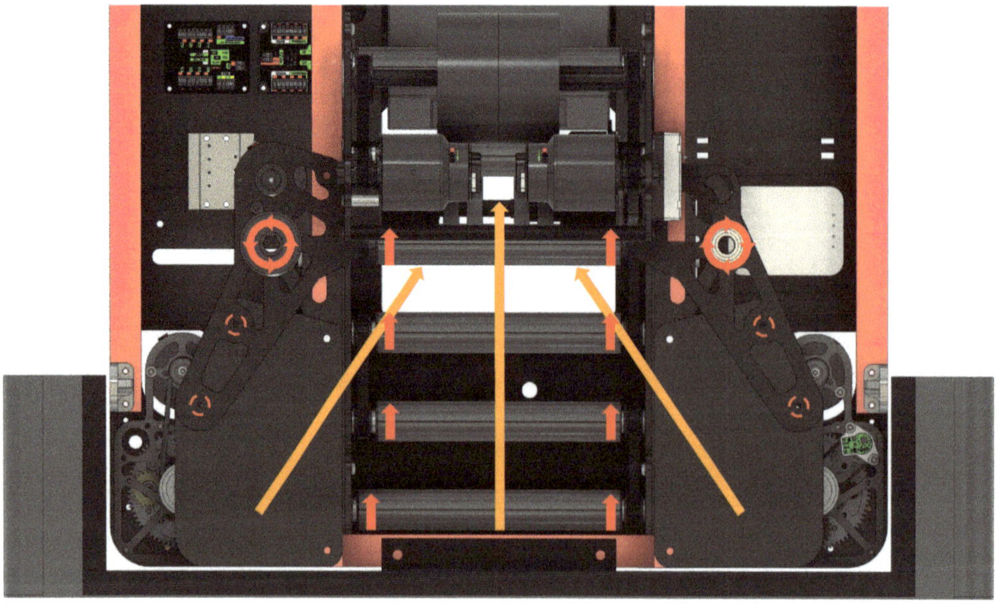

Diag. 4.14: Linear Indexer Ex. 3 Ball Path; Ref. 10

The momentum from the intake pushes the balls forward. If the balls are already centered, they simply follow the linear indexer (the direction in which the rollers spin is shown by all of the red arrows) along the center orange arrow. If the balls enter the indexer on the left or right side, they collide with structural pieces, highlighted by the dotted red circles, as they are hidden from view. This diverts their path such that the ball will eventually come into contact with the vertical rollers shown by the red circular arrows. This centers the ball before the ball reaches the robot's flywheel.

A single motor powers the entire indexer (both the linear indexer and the filter). The motor uses a combination of belts and bevel gears to power all the different elements.

There are additional ways to filter game objects before the game object reaches the launcher. This chapter is simply a guide that provides some examples, but there are many more creative ways to design an indexer not documented here.

## 4.6 Summary on Indexers & Storage Systems

Here's a quick review on the Indexers & Storage Systems:

**Linear Indexer**

- Simple design
- Limited storage
- Filtering Considerations (Intake ➢ indexer ➢ launcher)
    - Usually assumes objects filtered at intake
    - Can use sideways rollers to filter at beginning of indexer
- Game Object Path Considerations (intake ➢ indexer ➢ launcher)
    - Game Object enters usually through the side; can enter through the top
    - Game Object moves in a straight line
    - Game Object exits from the side; can be redirected to exit from the top/bottom with a feeder.

**Moving Floor Indexer**

- Complex design
- Very large storage
- Filtering Considerations (intake ➢ indexer ➢ launcher)
    - Filters towards corner of robot
- Game Object Path Considerations (intake ➢ indexer ➢ launcher)
    - Game Object enters through the top or side
    - Game Object moves linearly in one direction, then linearly perpendicular to the first direction
    - Game Object usually exits from the top with a feeder; can continue in the linear perpendicular direction to exit from the side

| Rotating Indexer |
|---|
| • Complex design<br>• Large storage<br>• Filtering Considerations (intake ➢ indexer ➢ launcher)<br>  ○ Filters towards anywhere on the edge of the circular path of the indexer<br>• Game Object Path Considerations (intake ➢ indexer ➢ launcher)<br>  ○ Game Object enters usually through the top; can enter through the side<br>  ○ Game Object moves in a circular motion<br>  ○ Centrifugal force feeds game objects into launcher; game object exits from the top with a feeder or from the side |

Diag. 4.15: Intake Summary Table

If looking at indexers in isolation, the following flowchart provides a way to analyze which indexer to use in a robot design. As previously noted, during robot design, often all mechanisms are being considered at the same time. In particular, decisions about the intake and launcher (flywheel/catapult) affect the indexer.

To reiterate, in the context of indexers, a feeder refers to an assembly that either pulls game objects out of an indexer and/or redirects the game object's direction of movement. This can be accomplished with rollers, wheels, belts with/without a wall (which is referred to as a backboard when combined with a wheel specifically).

Because filtering occurs along the game object's path through the robot, the choice of indexer is largely dependent on where that filtering is needed.

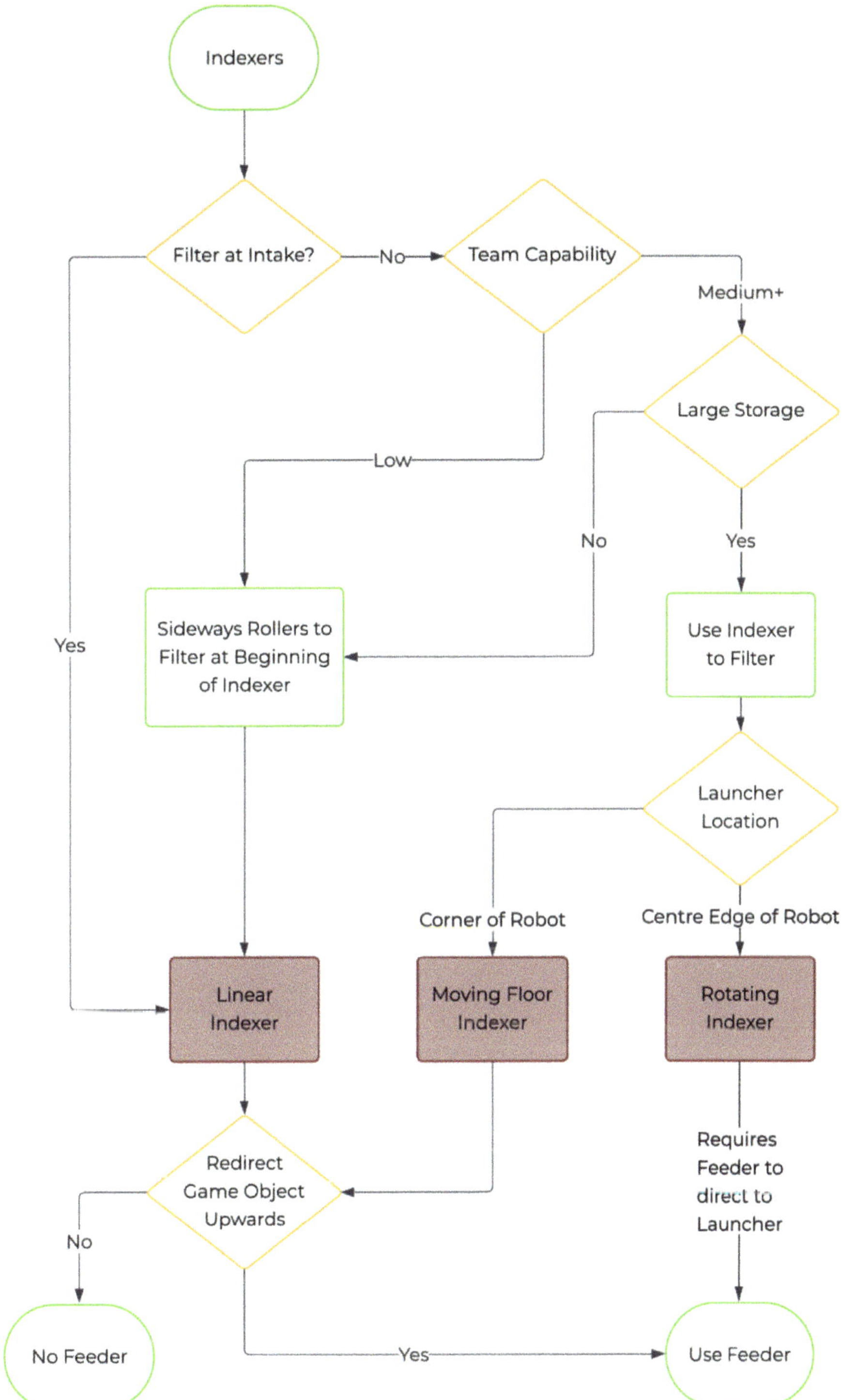


Diag. 4.16: Indexer Flowchart

# 5. Flywheels

## 5.1 Introduction

Flywheels are mechanisms that use spinning wheels to shoot game objects. Flywheels are one of two types of launchers, with the other being Catapults (see Chapter 6).

As noted in Subchapter 3.1 there is decent variability for game objects. There is not a set way that flywheels should be designed given this variability in game objects. This chapter provides examples of the different types of flywheels that have been used in previous FRC games to hopefully inspire new designs.

The design of a flywheel dictates the possible trajectories of the flywheel's shot.

Shot trajectories can be varied in a few different ways with flywheels. The simplest way is to vary release velocity of the flywheel by changing the speed the flywheel spins. A more complicated way to vary shot trajectory is by varying the release angle. In a vertical flywheel this is done by changing the shape of the hood in some way, while in a horizontal flywheel this is done by changing the angle of the ramp the flywheel sits on.

In general, shots become more predictable/accurate when spin is applied, as the angular momentum stabilizes its flight and prevents unwanted wobbling as the game object travels through the air. Think about top spin applied by tennis players in standard shots, and back spin applied by basketball players when shooting toward the hoop. This is known as gyroscopic stability.

An example of the difference between spin vs. no spin was demonstrated by Team 5987 and can be seen in online References 21 and 22 (refer to https://www.frc-handbook.com/ref/21 and https://www.frc-handbook.com/ref/22).

Spin can also change trajectories as will be covered when the different types of flywheels are discussed later.

That being said, spin is not necessary. A catapult does not create spin on the game object and can still land shots. Put simply, spin is generally beneficial if it can be achieved.

## 5.2 Single Vertical Flywheel

In a single vertical flywheel, a flywheel spins a game object, usually a ball, against a hood.

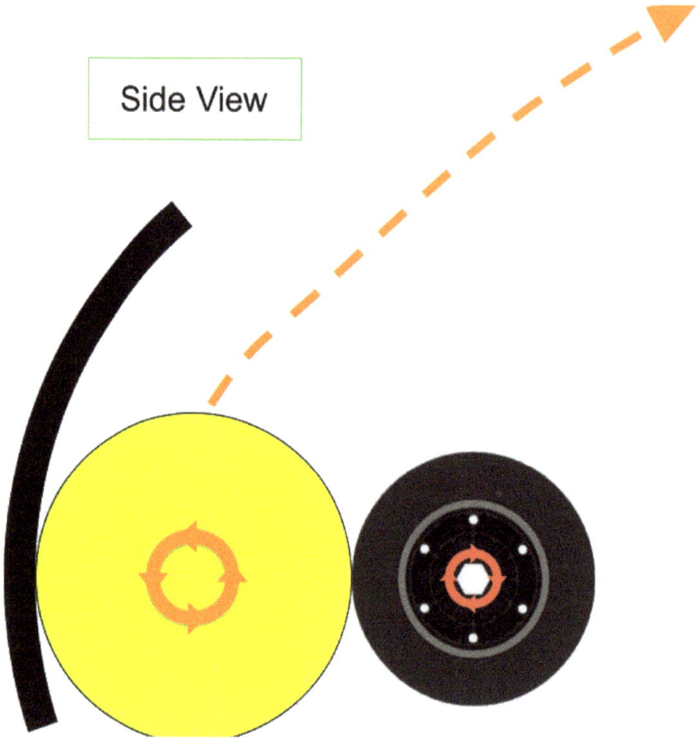


Diag. 5.1: Single Vertical Flywheel Side View Abstract

The black curved plate in diagram 5.1 represents the hood. The clockwise spinning wheel represents a flywheel, (direction depicted by red circular arrows). The flywheel naturally rolls the ball against the hood, creating a counterclockwise spin depicted by the orange circular arrows. This is known as backspin when the perspective is from behind the ball. The ball gets launched out at the right angle of a line drawn from the end of the hood to the center of the flywheel. Gravity then acts on the ball, causing the trajectory to curve downward as the ball continues moving, (shown by the dotted orange arrow).

In general, the longer the ball is in contact with the flywheel, the faster the ball's velocity is at release. This is because the flywheel is only accelerating the ball while it is in contact with the ball. One thought then would be to extend the length of the hood. This length could be extended in either direction. Extension below the flywheel changes how the indexer must feed the ball to the flywheel, (think straight up vs. sideways). Extension above the flywheel affects the trajectory angle of the flywheel mechanism. Adding too much length on this end makes certain types of shots impossible as the ball won't be able to be shot at a very high upward angle. As such, there is an optimal length for the hood depending on game mechanics.

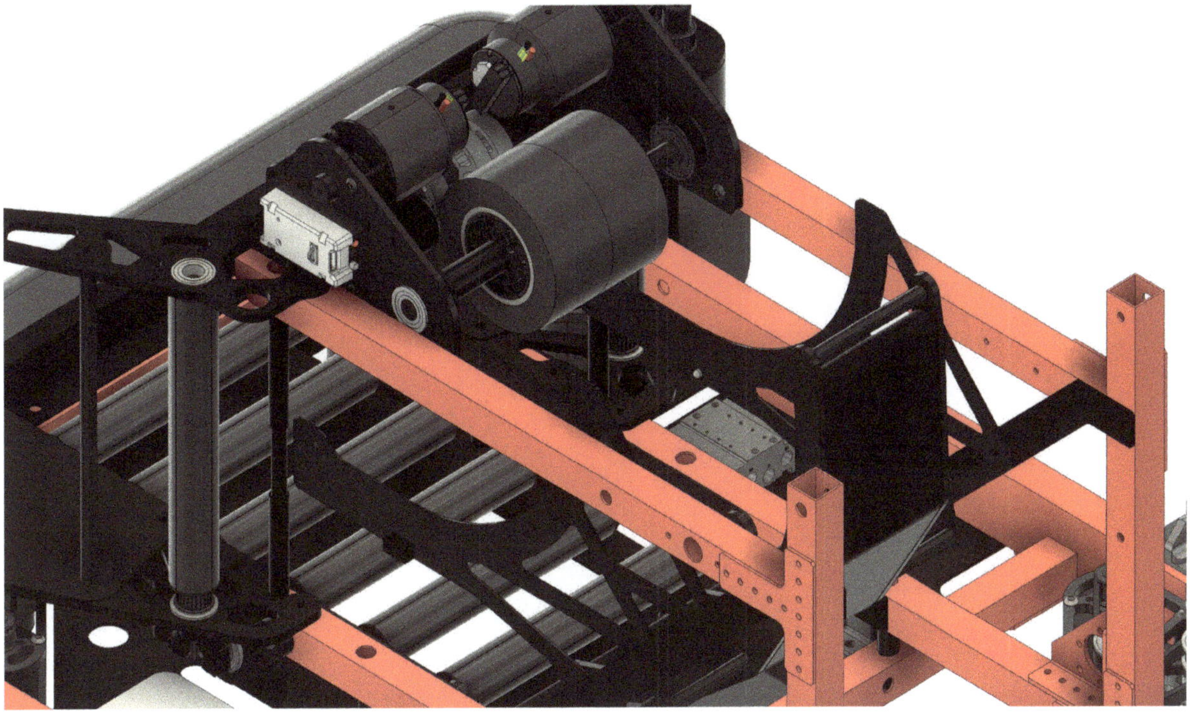

Diag. 5.2: Single Vertical Flywheel Ex. 1 Isometric View; Ref. 10

Diagram 5.2 is the first example of a single vertical flywheel. See https://www.frc-handbook.com/ref/10 for more information on the game and full robot.

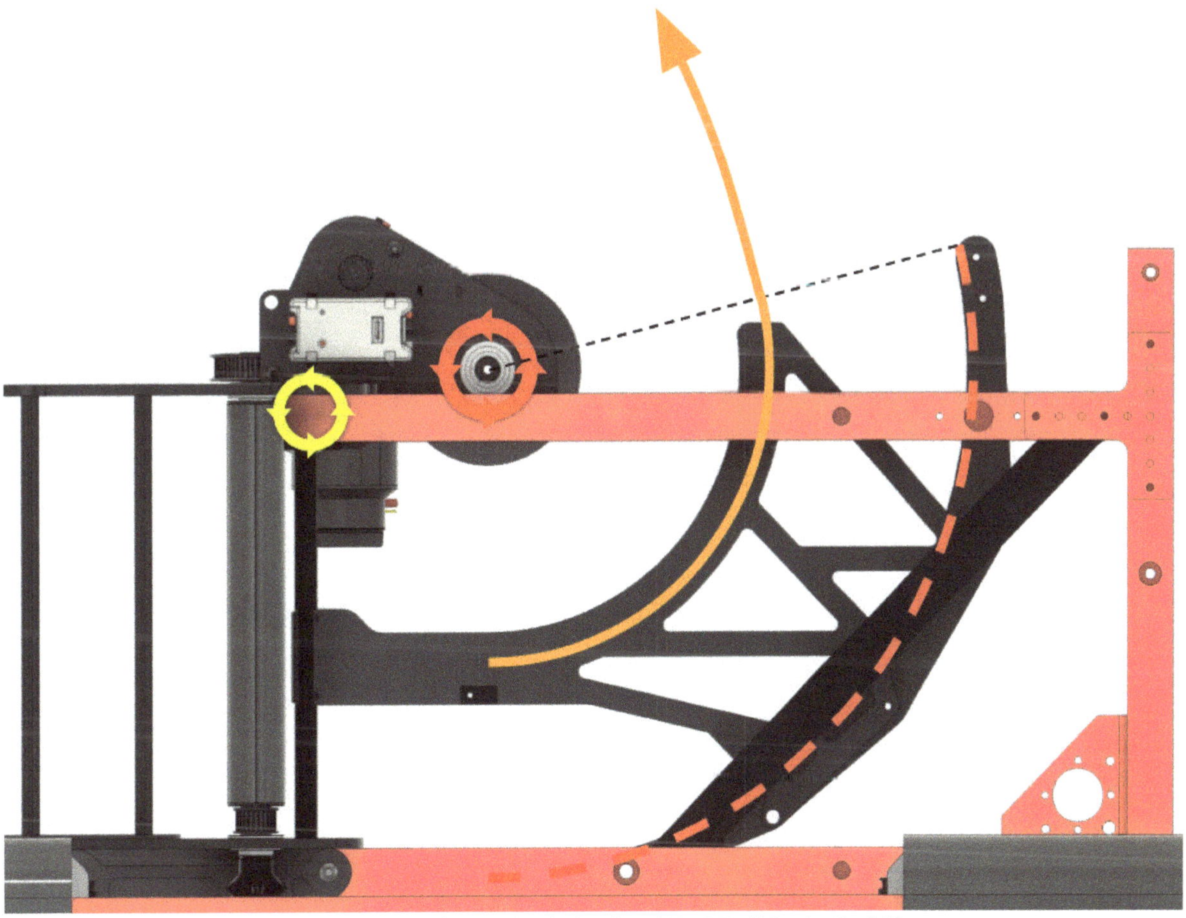

Diag. 5.3: Single Vertical Flywheel Ex. 1 Ball Path; Ref. 10

In diagram 5.3, a feeder from the indexer helps push the ball to the flywheel, as depicted by the yellow circular arrows. The dotted red lines outline the hood's surface that the ball rolls against. Given that this flywheel is designed to shoot the ball at a very steep upward angle while maximizing the contact of the flywheel with the ball (which in turn maximizes the release velocity of the flywheel), this hood makes sense. Most of the contact happens below the flywheel, with a very short section of the hood being above the flywheel. The flywheel's spin is depicted by the red circular arrows. The combination of flywheel and hood guides the ball's trajectory to roughly follow the orange arrow.

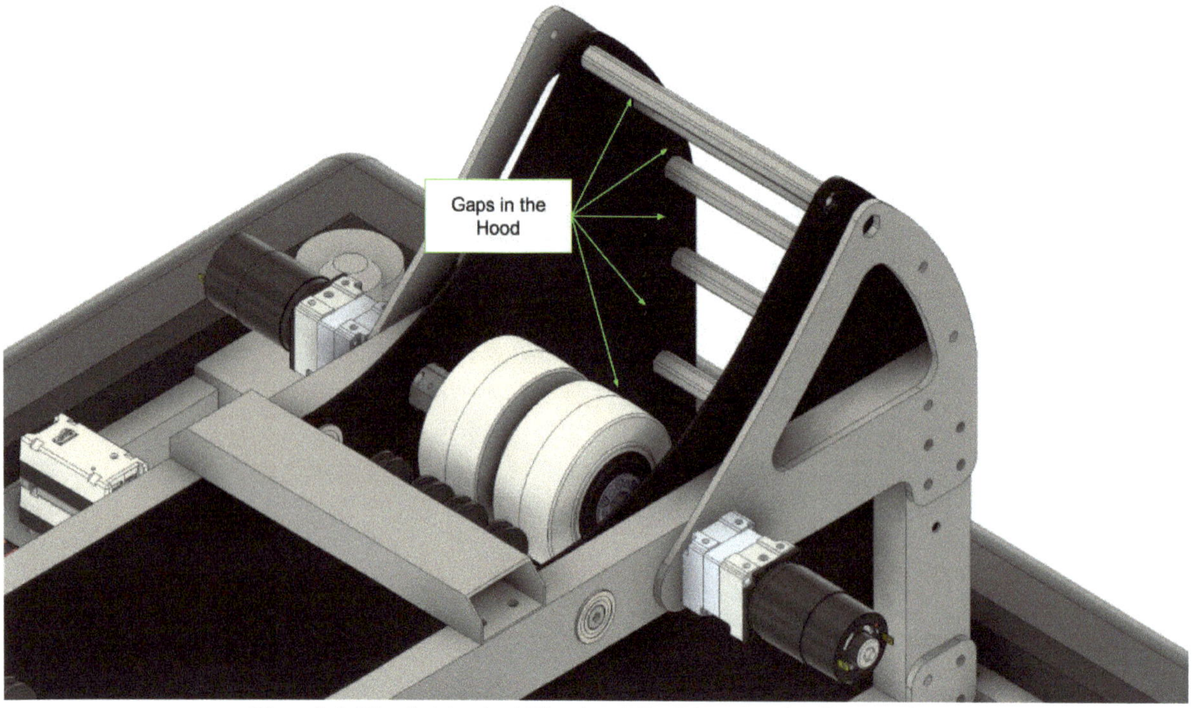


Diag. 5.4: Single Vertical Flywheel Ex. 2 Isometric View; Ref. 17

Diagram 5.4 is the second example of a single vertical flywheel. For more information on the game and the full robot see https://www.frc-handbook.com/ref/17.

Something to note about the design of hoods is that while it is possible to construct a single solid piece from a 3D printer or by bending a plate, it is also acceptable to have small gaps between many structural pieces going across the width of the flywheel mechanism. This results in a slightly less smooth flywheel, but the flywheel mechanism still functions.

This second flywheel example employs such a method for the construction of the hood of the flywheel. Due to the view being isometric, the angle may be deceiving, but the ThunderHex stock pieces running the width of the flywheel are equally spaced along the hood's path. The gap between each ThunderHex is roughly 2 inches and considered "small" or insignificant as the ball's size for the 2021 FRC game is 7 inches in diameter.

Apart from the difference in what the hoods are made out of, the macro design of the flywheels in Reference 10 and Reference 17 are somewhat similar.

Diagram 5.5 showcases another example of a single vertical flywheel from Team 1986. See details at https://www.frc-handbook.com/ref/19.

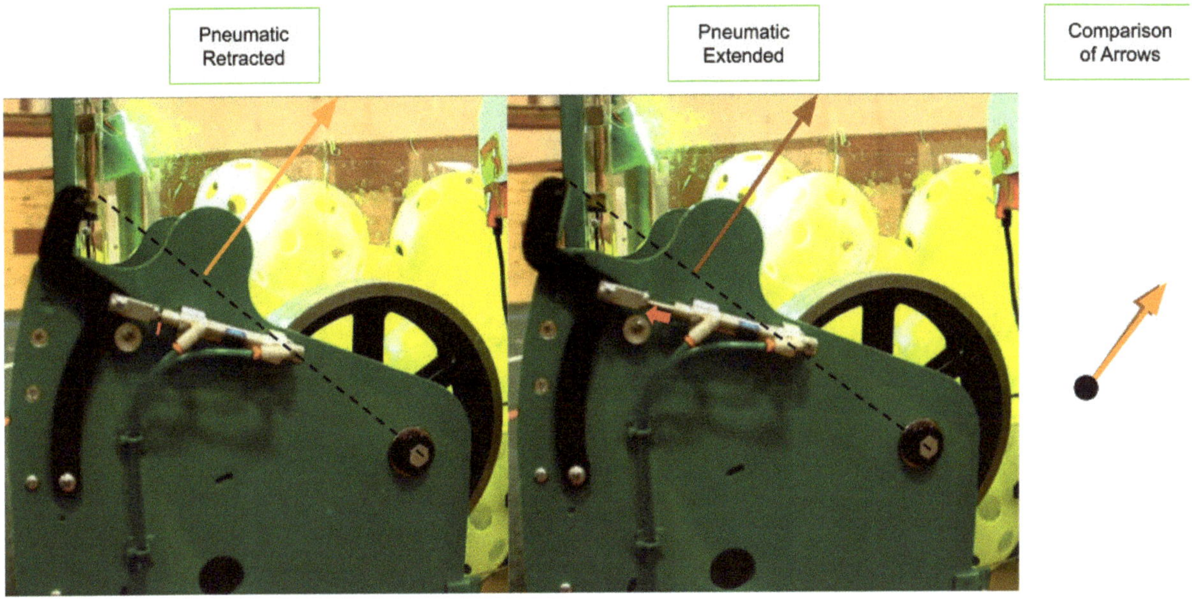


Diag. 5.5: Single Vertical Flywheel Ex. 3 Difference in Release Angle; Ref. 19

A notable feature of this flywheel is the adjustable hood of the flywheel. A pneumatic is either extended or retracted as depicted by the red line and arrow. This changes the trajectory angle of the flywheel slightly as shown by the comparison of the 2 different colored orange arrows. The angle adjustment of the flywheel may be slight, but for shots that travel as far of a distance as in the 2017 FRC game, such an angle adjustment allows the robot a lot more positional flexibility in the field to score balls. (Keep in mind that, it's not only the hood that is adjustable, but also the flywheel's speed.)

Another difference between this flywheel and the previous 2 examples is that the ball's path does not wrap around the wheel and instead has a very short window of contact with the flywheel. The ball is fed straight up to the flywheel by the indexer as opposed to sideways.

Reference 19 is one of many different ways to make an adjustable hood. Another way involves putting a section of the hood between bearings so it can be turned as if it were a small section of a very large gear.

Diag. 5.6: Single Vertical Flywheel Ex. 4 Isometric View; Ref. 9

While Reference 9 is not a fully completed CAD, the key parts of this adjustable hood can be seen. See https://www.frc-handbook.com/ref/9 for more information on the game and full robot.

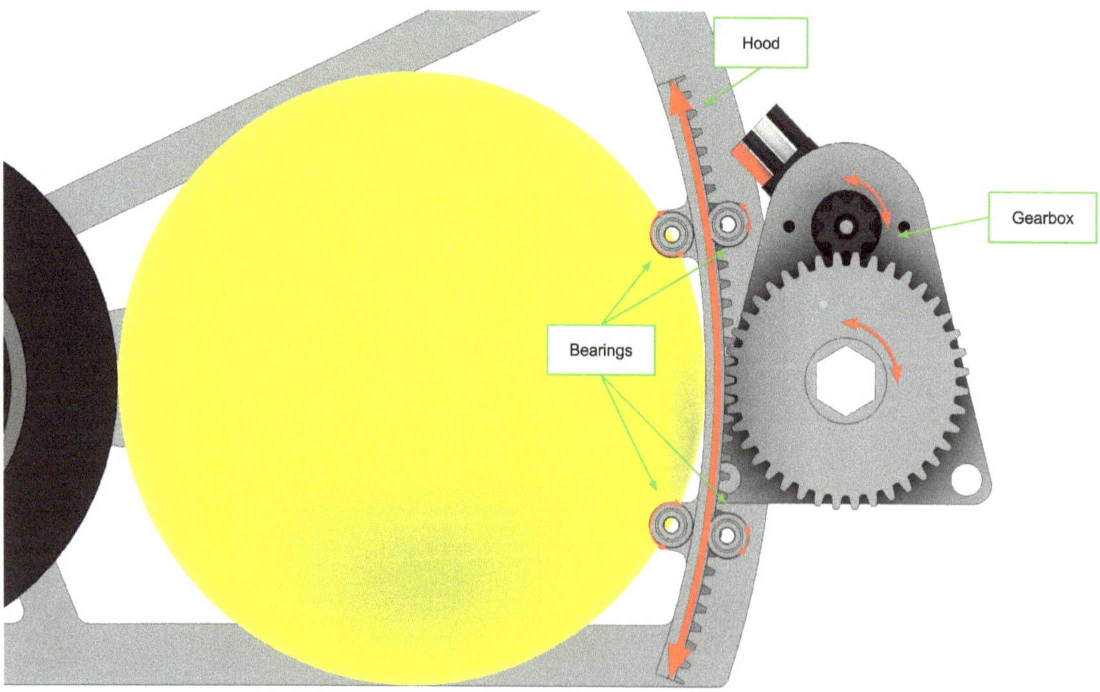


Diag. 5.7: Single Vertical Flywheel Ex. 4 Hood Movement; Ref. 9

As shown by all of the arcs of red arrows, the interlocking gears of this adjustable hood can spin in either direction to adjust the release angle of the flywheel. 4 bearings keep the hood piece in place. A gearbox drives the hood, allowing the hood to be adjustable.

## 5.3 Double Vertical Flywheel

A double vertical flywheel as a name actually refers to two different kinds of flywheel mechanisms. The first is the hooded double vertical flywheel. The second is the standard double vertical flywheel.

### Hooded Double Vertical Flywheel

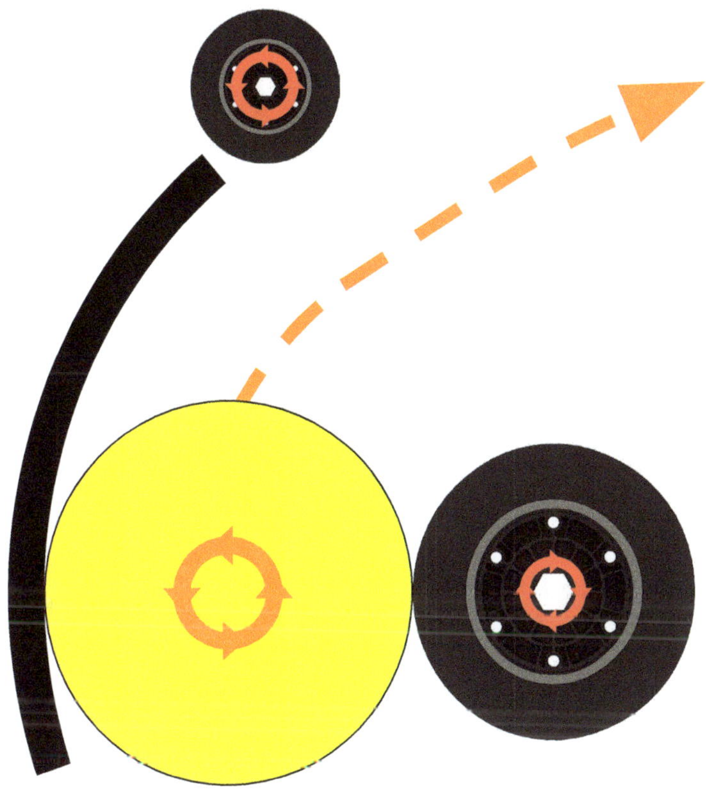

Diag. 5.8: Hooded Double Vertical Flywheel Side View Abstract

How does a hooded double vertical flywheel work? The abstract design may look familiar because it is similar to diagram 5.1. This version of a double vertical flywheel is literally a single vertical flywheel with an additional smaller flywheel added at the end of the hood. (See Subchapter 5.2 on Single Vertical Flywheel).

The additional smaller secondary flywheel spins in the opposite direction of the larger primary flywheel as indicated by the red circular arrows. The purpose of the additional flywheel is usually to increase the range of a flywheel mechanism and can be thought of as a direct upgrade over the typical single vertical flywheel. The additional flywheel further accelerates the ball as the ball is being released by the flywheel mechanism. Spin is still created in this type of flywheel design. The ball is in contact with the primary flywheel and hood as it would be in a single vertical flywheel. Additionally, while the secondary flywheel does increase the acceleration of the ball, assuming the two flywheels are running at the same RPM, the surface speed of the smaller flywheel is less than the larger flywheel given the difference in circumference. (A way to think about it is that for each rotation, the larger flywheel covers more "distance" than the smaller flywheel.) This results in the side of the ball closer to the primary flywheel being accelerated more than the side closer to the secondary flywheel, inducing further spin on the ball. This spin creates gyroscopic stability, producing a more predictable and consistent shot.

Refer to https://www.frc-handbook.com/ref/20 for an example of a hooded double vertical flywheel by Team 254.

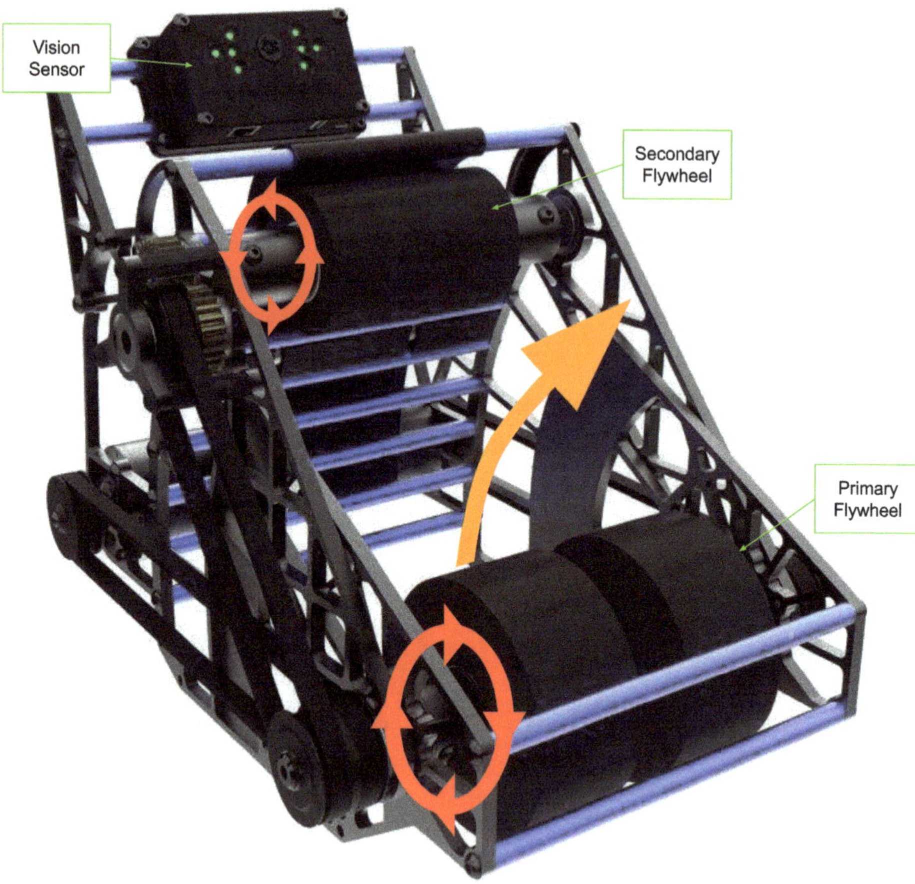


Diag. 5.9: Hooded Double Vertical Flywheel Ex. 1 Ball Path; Ref. 20

This flywheel requires the ball to be fed from directly underneath by the indexer. The primary flywheel drives the ball up the flywheel highlighted by the larger red circular arrows. Then the primary flywheel and secondary flywheel, (highlighted by the smaller red circular arrows) work in conjunction to release the ball. This directs the ball to follow the path of the orange arrow.

It can be noted that there is not a solid hood and instead multiple stock pieces running across the width of the flywheel. This is because the acceleration from the secondary flywheel makes a solid hood unnecessary.

A special feature of this flywheel is that there is a vision sensor positioned above the secondary flywheel. Paired with good programming, this allows the robot to autonomously map its location on the field and adjust the direction and speed of the flywheels to score shots.

### Standard Double Vertical Flywheel

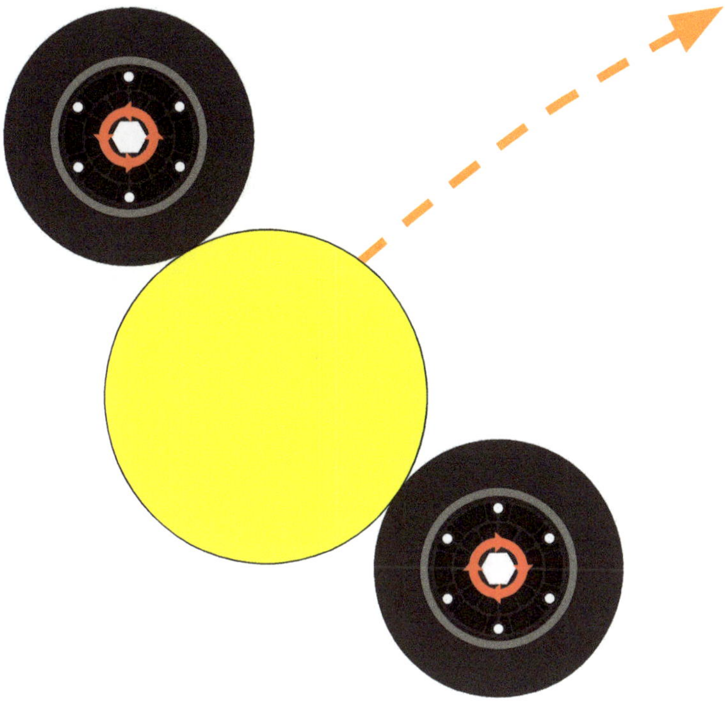

Diag. 5.10: Standard Double Vertical Flywheel Side View Abstract

Moving on to the standard double vertical flywheel depicted in diagram 5.10, two flywheels of equal size spin and make brief contact with the ball before shooting the ball out. Spin of the ball can be induced by varying the speeds of the individual flywheels, creating gyroscopic stability. However, if both flywheels rotate at the same speed, the ball will be thrown out of the robot without spin as depicted in diagram 5.10. Trajectories can, of course, be varied by the speed of the flywheels, with faster speeds accelerating the ball more, leading to longer shots. This type of double vertical flywheel design will be referred to as the standard double vertical flywheel from here on.

Over the years Team 4817 has used variations of the standard double vertical flywheel design to great effect.

Diag. 5.11: Standard Double Vertical Flywheel Ex. 1 Isometric View; Ref. 11

This example in diagram 5.11 shows a pretty straightforward standard double vertical flywheel. See https://www.frc-handbook.com/ref/11 for more information on the game and full robot.

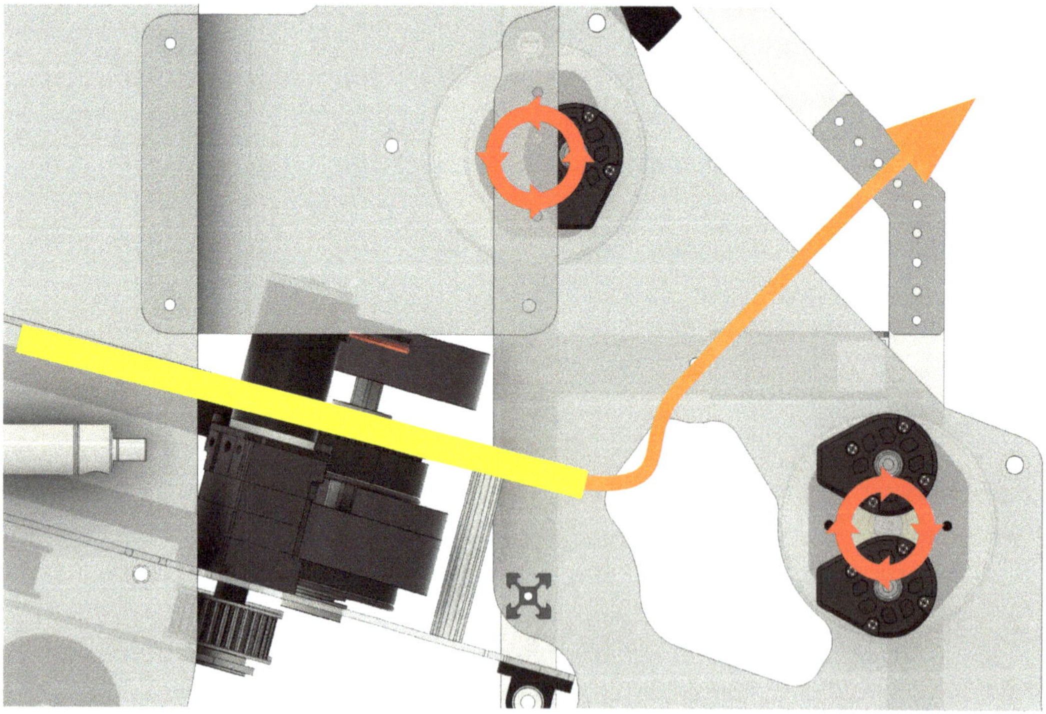

Diag. 5.12: Standard Double Vertical Flywheel Ex. 1 Ball Path; Ref. 11

In diagram 5.12, the flywheels spin in the direction indicated by the red circular arrows. Propelled by the indexer, the ball rolls past the top left flywheel without contacting it, shown by the yellow section of the ball path line. However, once the ball rolls forward far enough, the ball begins to contact the bottom right flywheel. This pushes the ball up and towards the top left flywheel. Then both flywheels work in conjunction to launch the ball.

Two 775pro motors were used for the top left flywheel, which would be roughly equivalent to 1 modern FRC motor like a Falcon 500. Four 775pro motors were used for the bottom right flywheel, also roughly equivalent to 2 modern FRC motors. By making the bottom right flywheel stronger, back spin is induced by the flywheel on the ball as it leaves the robot.

The following example is a variation of the standard double vertical flywheel. See https://www.frc-handbook.com/ref/8 for more information on the game and full robot.

Diag. 5.13: Standard Double Vertical Flywheel Ex. 2 Isometric View; Ref. 8

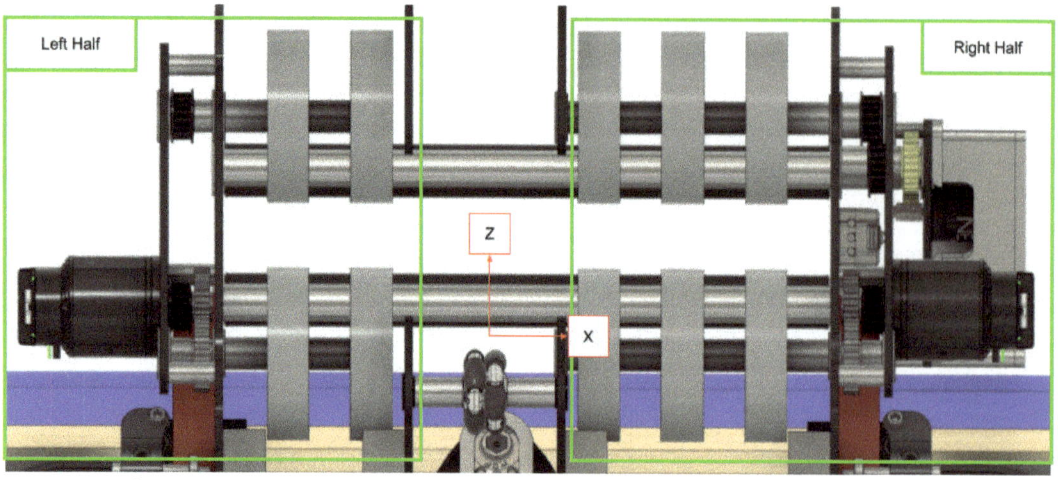


Diag. 5.14: Standard Double Vertical Flywheel Ex. 2 Top Front View; Ref. 8

The first thing to note about this flywheel is that it is split vertically roughly in half as can be seen from the front view in diagram 5.14. Due to the game object being a ring, it was determined the way to achieve gyroscopic stability was to induce spin around the z axis instead of the typical x axis as it would be for a ball. To achieve this, one side of the ring has to be accelerated more than the other side. By splitting the flywheel in half like this, the left half could be spun at a lower speed than the right half, inducing a spin on the ring. This typically creates the problem of a sideways curved trajectory as is further described in Subchapter 5.4. However, given how the game was designed, shots were generally taken from a relatively short distance, so this sideways curving trajectory did not really affect how the robot aligned itself to the Speaker goal.

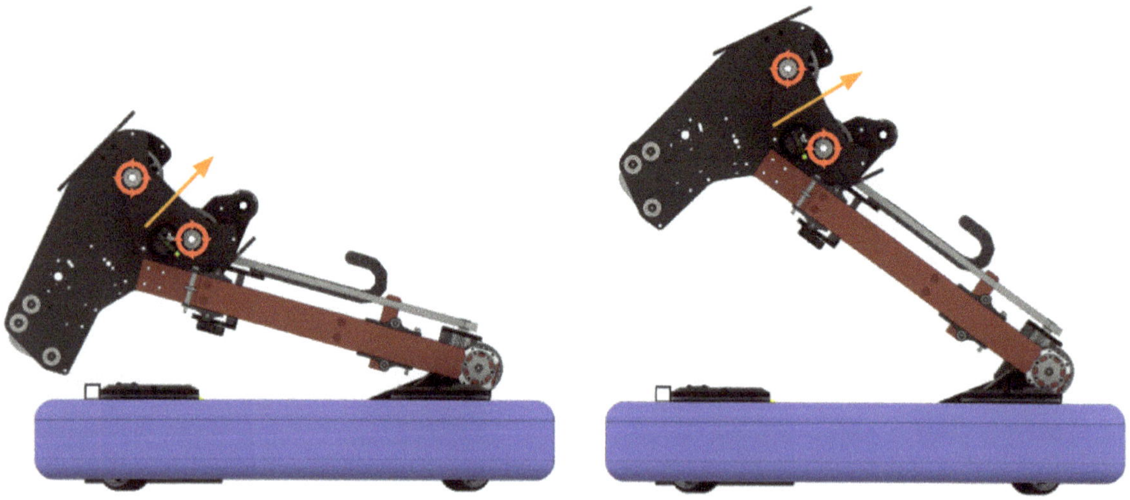

Diag. 5.15: Standard Double Vertical Flywheel Ex. 2 Ball Path; Ref. 8

This robot does not feature an indexer mechanism. Instead, the ring is fed directly to the flywheel from the intake. The two flywheels spin in opposite directions to launch the ring. A single motor powers each half of this flywheel design. Because the top and bottom wheels are powered by the same motor in each half of this flywheel, the shot angle can only be adjusted by turning the extendable arm mechanism (see Subchapter 9.3 on Extendable Arms) that the flywheel assembly sits on.

To be clear, usually the top and bottom flywheels are powered by their own motor as described for Reference 10. Reference 7 is a more unique example to show how a general idea of a mechanism can be modified to work for a specific game.

## 5.4 Single Horizontal Flywheel

In horizontal flywheels, the spinning flywheels are placed to the left and/or right of the game object.

For a single horizontal flywheel, a wheel is on one side while a wall is on the opposite side. The wall and flywheel are put on a ramp so that the flywheel is angled upward, allowing game objects to fly in an arc.

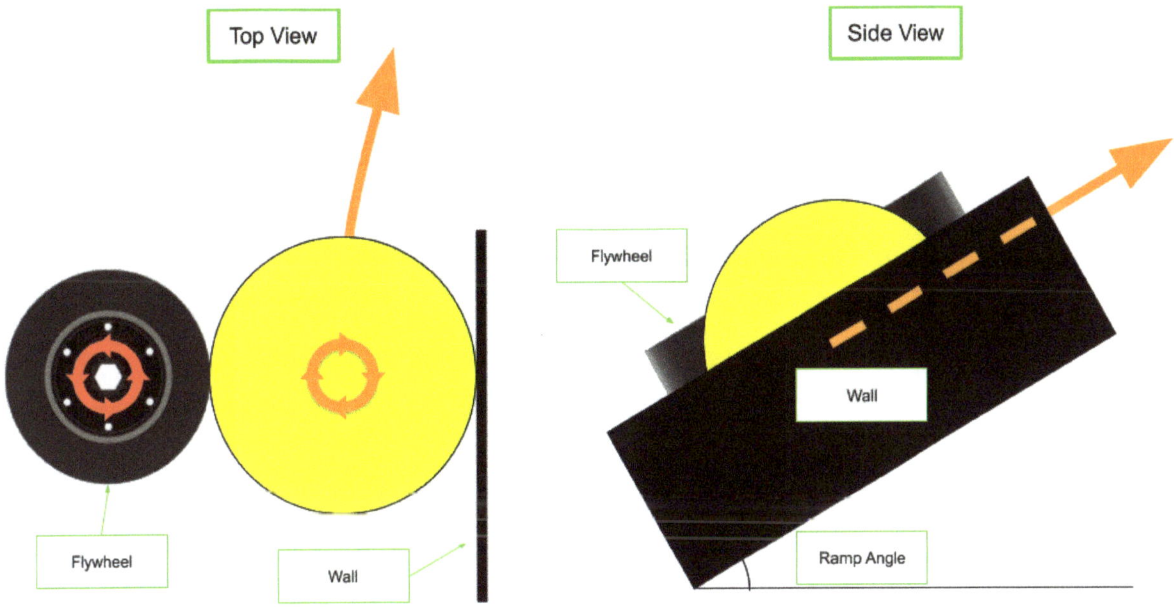


Diag. 5.16: Single Horizontal Flywheel Abstract

The structure of a horizontal flywheel naturally imparts a horizontal spin on the game object, and while this creates gyroscopic stability, shots over long distances will be affected by a noticeable magnus effect, which results in the trajectory of a game object curving to the side. This curve is exaggerated by the orange ball path arrow as depicted in the top view in diagram 5.18. While this trajectory is usually predictable and consistent, it can be awkward to have to aim to the side of a target to actually land shots over a long distance since this is less intuitive. This is one of the reasons why in general vertical flywheels are preferred over horizontal flywheels.

To vary shot trajectories, the flywheels can be spun at varying speeds. The other way to adjust the trajectory of the flywheel is by adding a pivot mechanism that changes the ramp angle of the horizontal flywheel. While not a horizontal flywheel, Reference 8 provides an example of a pivot adjusting the ramp angle of the flywheel (see Subchapter 5.3 for more details).

Refer to https://www.frc-handbook.com/ref/14 for a RI3D example from University of Minnesota Robotics.

Diag. 5.17: Single Horizontal Flywheel Ex. 1 Ball Path; Ref. 14

This robot utilises the standard design of a horizontal flywheel as shown in the previous abstract diagram.

A notable thing about this design is that the flywheel is moved to the right quite a bit due to the need to deform the ring to ensure good contact given the ring's properties.

## 5.5 Double Horizontal Flywheel

A double horizontal flywheel is very similar to a single horizontal flywheel. Instead of a wall on one of the sides, flywheels are placed at both sides.

It is possible to run both flywheels at the same speed to shoot a game object, however this results in inconsistent shots as gyroscopic stability is not achieved (see Subchapter 5.1's comparison of Reference 21 and Reference 22).

Compare diagram 5.18 with diagram 5.16. The wall on the right has been replaced with a flywheel. In a double horizontal flywheel, usually, one of the flywheels will spin a bit slower to create spin on the game object to achieve gyroscopic stability. In this case, since the spin is still horizontal, all of the same issues of a single horizontal flywheel's trajectory still apply.

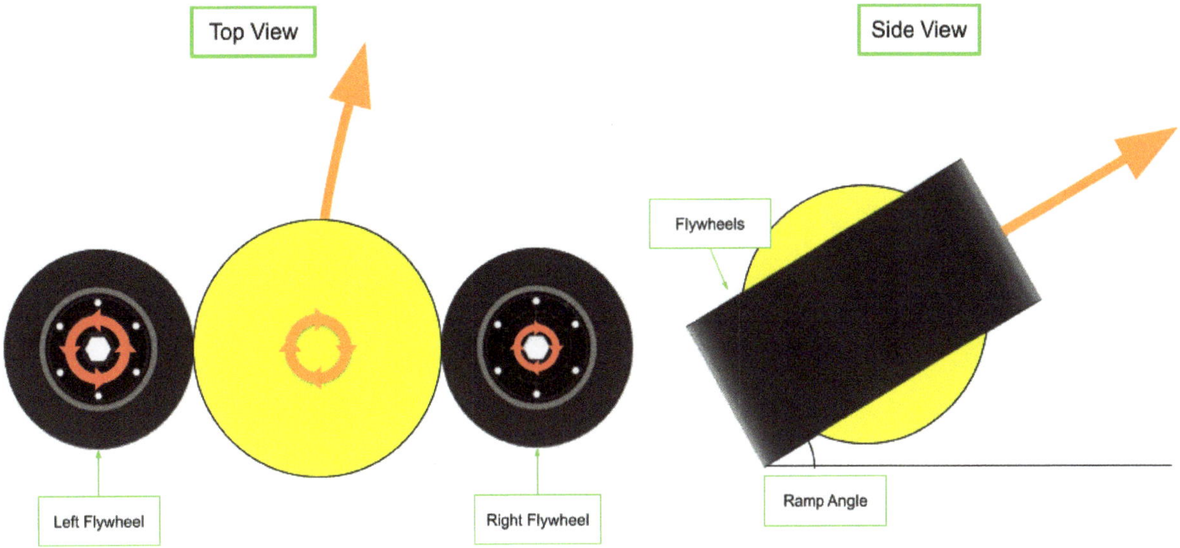


Diag. 5.18: Double Horizontal Flywheel Abstract

The difference between a double horizontal flywheel and a single horizontal flywheel is that there are now 2 wheels pushing the game object instead of one. While not necessarily true, this usually means more motors are involved in driving the flywheel mechanism, which can enable longer distance shots.

There are very few actual FRC examples of a double horizontal flywheel. It would appear most teams have made the determination that this isn't the best type of flywheel for recent FRC games. That being said, footage of a double horizontal flywheel prototype can be found at https://www.frc-handbook.com/ref/21. This prototype was developed by Team 5987 before they settled on a double vertical flywheel shooter for the 2024 Game.

## 5.6 Turntables (for Flywheels)

A turntable—a rotatable platform that supports another mechanism—allows a flywheel to pivot independently of the drivebase, enabling a robot to shoot in any direction regardless of the orientation of the drivebase. While it is possible to simply turn the drivebase to align for shots, a turntable improves efficiency and grants additional flexibility for how a robot can play a game. A turntable, coupled with april tag tracking and sensors, enables automatic alignment of the flywheel with game goals. Turntables can also be used in a similar way for other types of mechanisms like lifts and arms if those mechanisms need to have the ability to rotate relative to the robot's base.

For flywheel designs that involve feeding balls from directly underneath the flywheel, it is easy to add a turntable. No matter which direction the flywheel is pointing, game objects can still be fed to the flywheel from below. As such, turntables are primarily used with hooded flywheel designs where a game object is fed to the flywheel from directly under.

On the next pages are some examples of turntable designs.

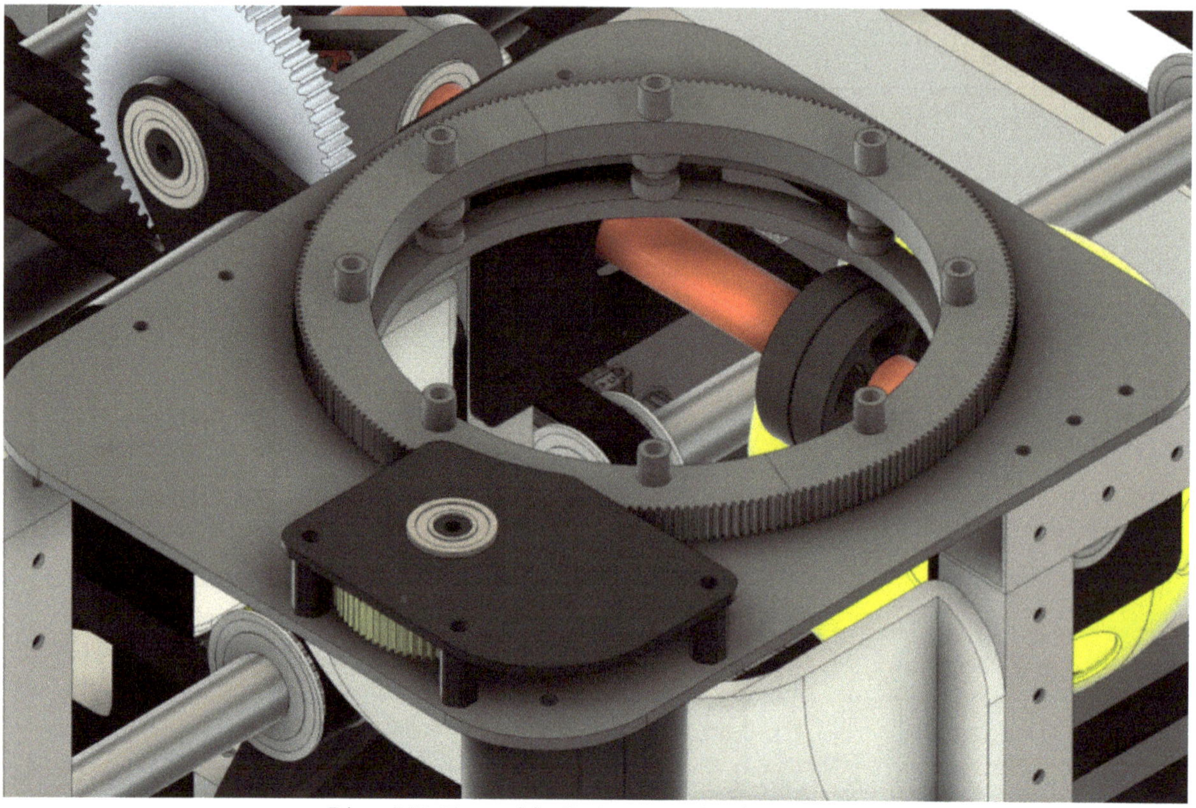

Diag. 5.19: Turntable Ex. 1 Isometric View; Ref. 18

The first turntable example is from Team 4817. (For more information on the game and full robot see https://www.frc-handbook.com/ref/18.)

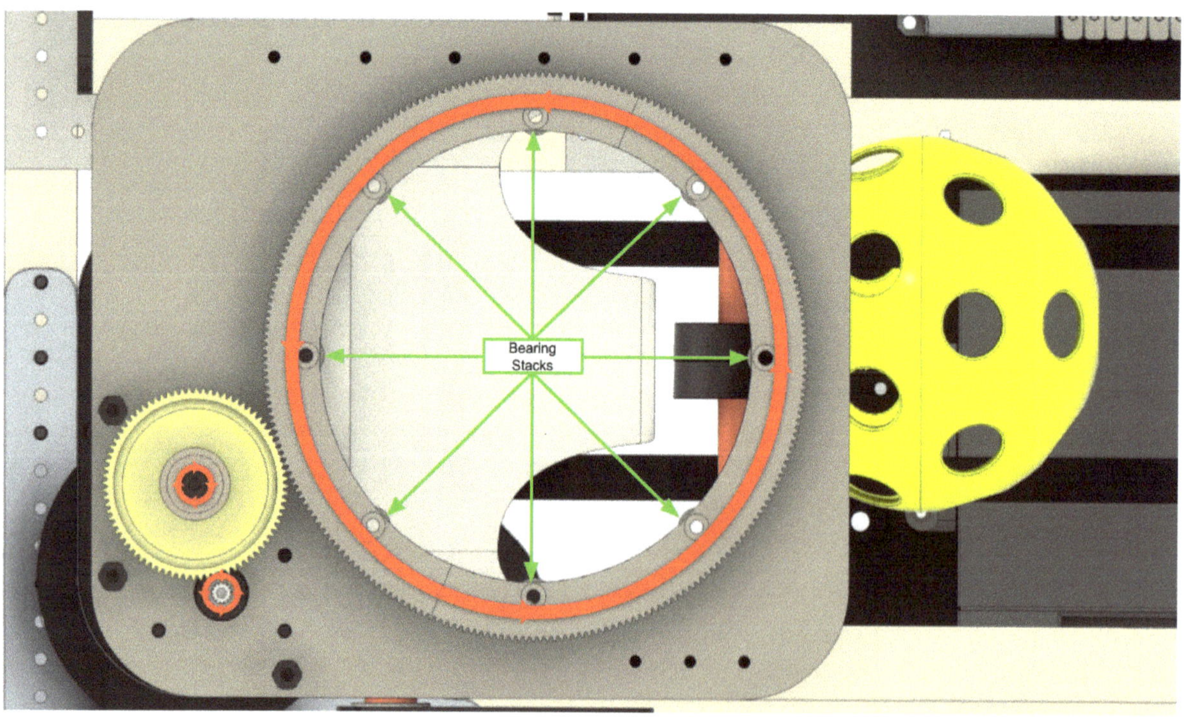


Diag. 5.20: Turntable Ex. 1 Top View; Ref. 18

This turntable works by powering a motor to turn a series of gears, with the final driven gear integrated into the turntable.

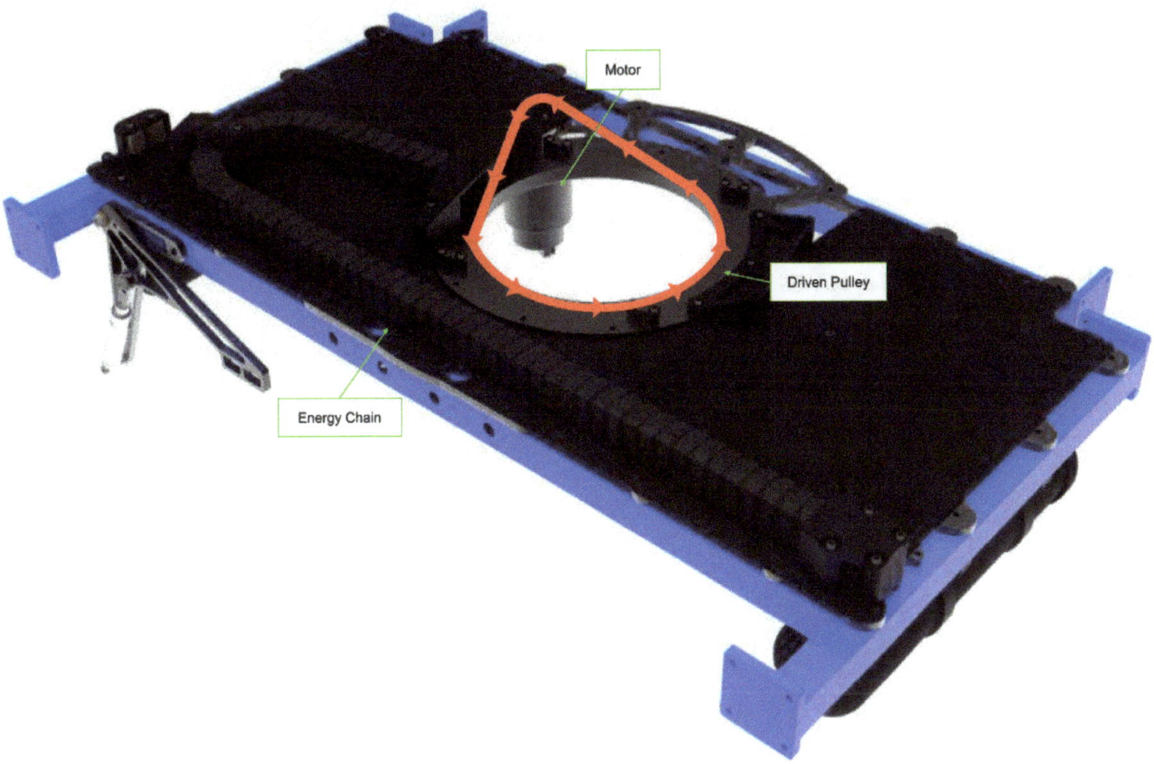


Diag. 5.21: Turntable Ex. 2 Isometric View; Ref. 20

Diagram 5.21 shows another example of a turntable—this one designed by Team 254. This turntable is driven by a pulley and belt system, hidden from view, with the motor and driven pulley positions labeled.

This turntable is superior to the other turntable example as it includes an effective way to address wire management via the use of an energy chain. Otherwise wires would be unguided and may stick out in any direction.

## 5.7 Further Comparison of Flywheels

Due to the sideways curving of projectile trajectories, horizontal flywheels are rarely the optimal design choice. Typically designing a flywheel means choosing some type of vertical flywheel design: a single vertical flywheel, a hooded double vertical flywheel, or a standard double vertical flywheel.

The first comparison to make is between the two hooded types of vertical flywheels, the single vertical flywheel and the hooded double vertical flywheel. The added benefits of having the additional wheel of the hooded double vertical flywheel were already described in the first section of Subchapter 5.2. One reason against making a single vertical flywheel into a hooded double vertical flywheel would be the added complexity of the additional wheel. Another reason would be that the extra range capabilities of the design may not be necessary for the game. If it is not necessary to shoot very far, or it can't be done with enough accuracy, the extra range benefit is useless and complicates the design.

With that settled, let's compare the standard double vertical flywheel to the hooded options. The reason to use a double flywheel design over the hooded flywheel options is either for simplicity's sake or due to space constraints.

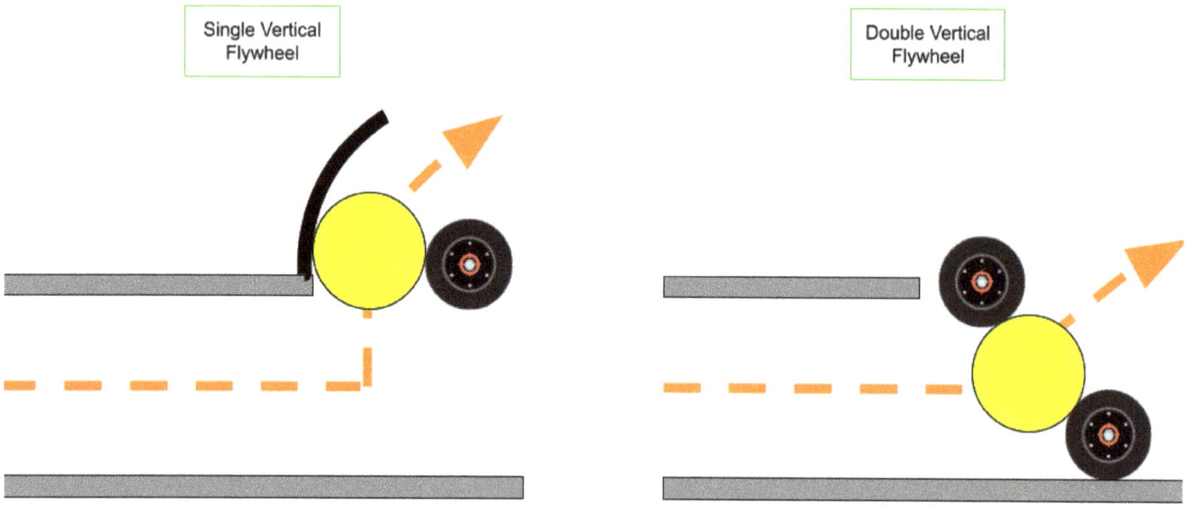


Diag. 5.22: Single Vertical Flywheel vs. Double Vertical Flywheel Side View Comparison Abstract

A standard double vertical flywheel that intends to shoot forward is more compact than a hooded flywheel in the sense that the ball would have to first travel upward before reaching a hooded flywheel. (To be absolutely clear, a compact single vertical flywheel is only possible if the ball is being shot the direction that it gets intaked from. An example of such a robot would be Reference 10.) Another factor is that it is often easier to design a smooth transition from an indexer to a standard double vertical flywheel than to a hooded flywheel. Creating an indexer that effectively pushes a ball from sideways to suddenly straight up into a hooded flywheel is not as simple as one might think.

## 5.8 Summary on Flywheels

Below is a summary of the various flywheels discussed. Note the table reflects what is generally true for all mechanisms of said type, however additional innovation is always possible to further improve the capabilities of a mechanism. For example, Reference 8 showcases a split double vertical flywheel design that is able to create spin around the flywheel's z axis as opposed to the usual x axis.

<table>
<tr><th colspan="3">Flywheels</th></tr>
<tr><td colspan="3"><ul><li>Trajectory controlled by speed of flywheel(s), spin of game object, and angle of release</li><li>Application of spin improves accuracy with top preferred as side spin causes game object's trajectory to curve sideways</li><li>Applicable for most FRC game objects</li></ul></td></tr>
<tr><td rowspan="3">Vertical Flywheel</td><td>Single Vertical Flywheel</td><td><ul><li>Game object spun between a hood and large flywheel</li><li>Back spin generated</li><li>Longer contact means greater velocity</li><li>Angle of release affected by hood angle</li><li>Hood angle can be made adjustable with a pivot arm or linear slide</li></ul></td></tr>
<tr><td>Hooded Double Vertical Flywheel</td><td><ul><li>Game object spun between a hood and a large flywheel; hood ends with a small flywheel</li><li>Back spin generated</li><li>Longer contact means greater velocity</li><li>Angle of release affected by hood angle</li><li>Hood angle can be made adjustable with a pivot arm or linear slide</li></ul></td></tr>
<tr><td>Standard Double Vertical Flywheel</td><td><ul><li>Game object contacts 2 identical large flywheels</li><li>Top/Back spin created if flywheels spun at different speeds</li><li>Angle of release affected by 2 flywheels positioning</li></ul></td></tr>
<tr><td rowspan="2">Horizontal Flywheel</td><td>Single Horizontal Flywheel</td><td><ul><li>Game object spun between side wall and large flywheel</li><li>Side spin</li><li>Flywheel positioned to left or right of game object</li><li>Angle of release affected by ramp angle</li><li>Ramp angle can be made adjustable with pivot arm</li></ul></td></tr>
<tr><td>Double Horizontal Flywheel</td><td><ul><li>Game object spun between 2 identical large flywheels.</li><li>Side spin created if flywheels spun at different speeds</li><li>Flywheel positioned on left and right of game object</li><li>Angle of release affected by ramp angle</li><li>Ramp angle can be made adjustable with pivot arm</li></ul></td></tr>
<tr><th colspan="3">Turntables</th></tr>
<tr><td colspan="3"><ul><li>Supplementary mechanism to orientate flywheels</li></ul></td></tr>
</table>

Diag. 5.23: Flywheel Summary Table

A flowchart summarizing the decision making process for the various flywheel designs can be found below.

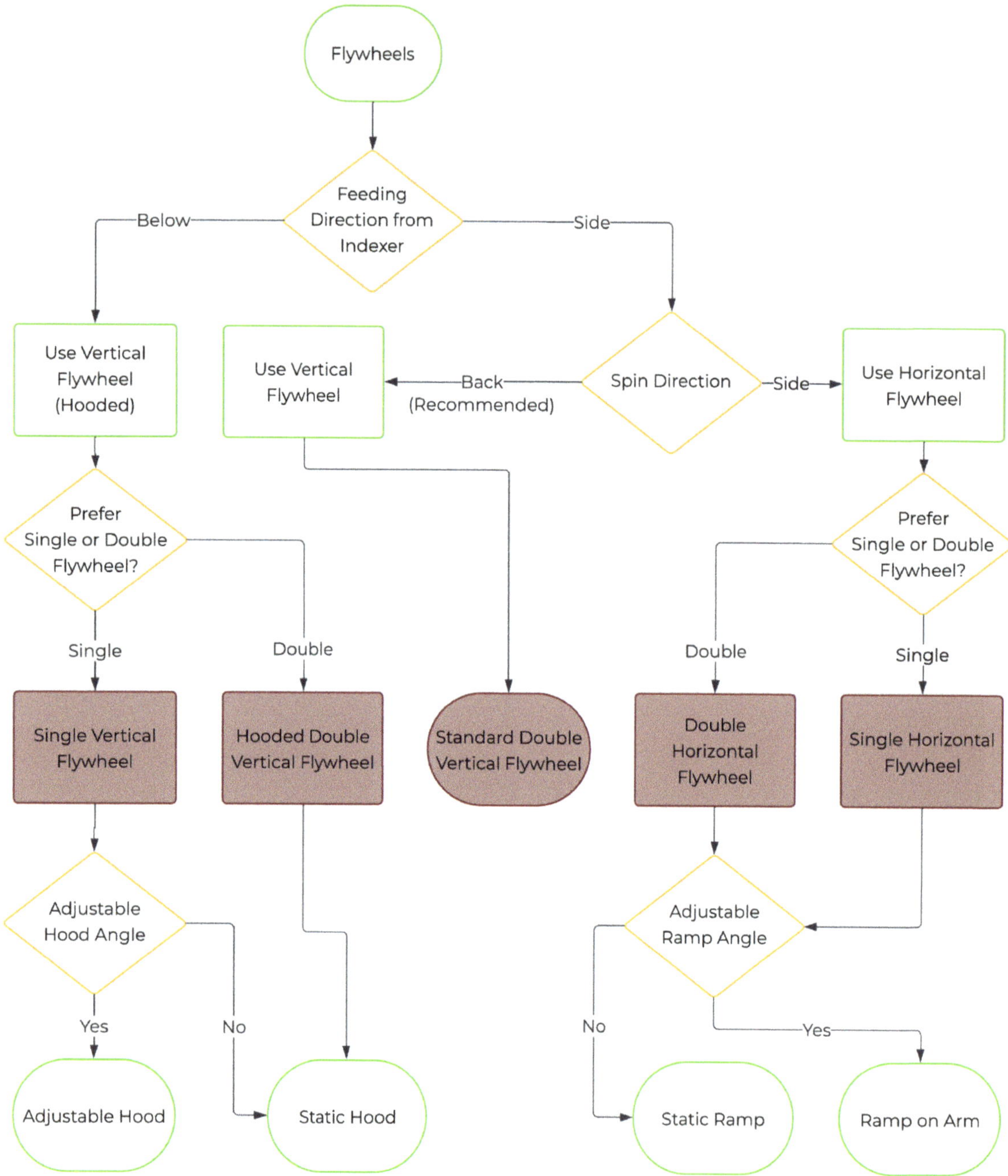


Diag. 5.24: Flywheel Flowchart

# 6. Catapults

## 6.1 Introduction

Most launchers used in FRC are flywheels. However, when a game object is particularly heavy, it may make sense to use a catapult mechanism to launch the game object. This chapter is about the various types of catapult designs and how they work.

A catapult works by storing up potential energy and then suddenly releasing that energy to swing a pivot arm to launch a projectile. Due to its ability to store up a lot of potential energy, catapults may be favored over flywheels when launching heavy game objects. There are a few different mechanisms that catapults use to achieve this.

First, the source of potential energy. Usually a mechanism that can store elastic potential energy is used, like springs or rubber cords. In theory, other sources of potential energy could come from gravitational or electrical energy, but these are not as straightforward to implement.

Second, the storing and releasing of the source of potential energy. There are a few different release mechanism options: different types of cams, sheared gears, a quick release winch, or pneumatics. Each option will be explained with an abstract diagram.

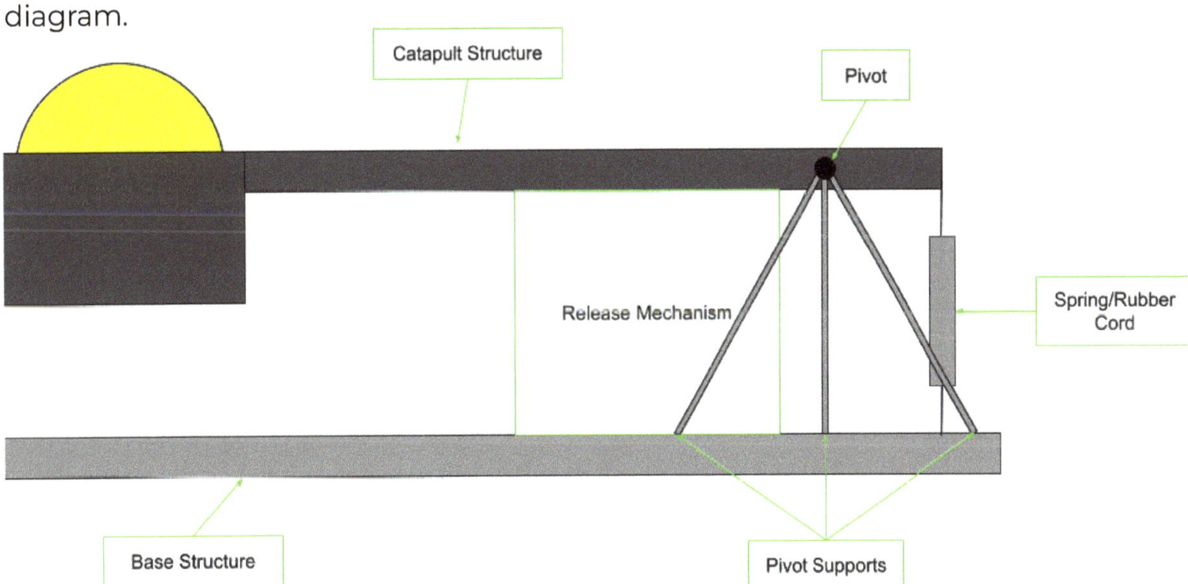


Diag. 6.1: Catapult General Abstract

Regardless of the release mechanism, a catapult always has these components: a base structure, a pivot and a mechanism that stores potential energy. The base structure is where the catapult's pivot sits atop of. Typically this would be the robot's drivebase, though not necessarily. The pivot is supported by a triangular structure as the triangle is considered the strongest and most stable shape in engineering. The catapult's structure is designed to rotate around the pivot. The mechanism that is the source of potential energy is on one side of the pivot and the release mechanism is usually on the other. Diagram 6.1 is a good representation. This general framework allows the catapult to store potential energy as the release mechanism pulls the catapult's arm downward. Then all of that potential energy is directed into launching the projectile when the release mechanism is suddenly triggered.

Given all catapults work this way, the following subchapters will focus on the different release systems.

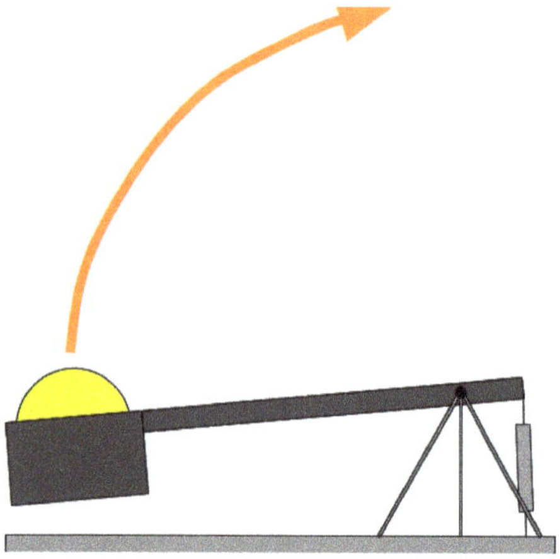

Diag. 6.2: Catapult Trajectory Abstract

As a general point, a catapult's game object trajectory follows an arc.

While most designs will not allow for this, the trajectory arc can be adjusted if the catapult is designed to vary the amount of potential energy stored before shooting the game object.

It can also be noted from diagram 6.1 that turning the catapult slightly may store enough potential energy to launch the game object. That is dependent on the elasticity of the spring or rubber cords.

In the next subchapters, the different catapult release systems are discussed.

## 6.2 Choo-Choo CAM Catapults

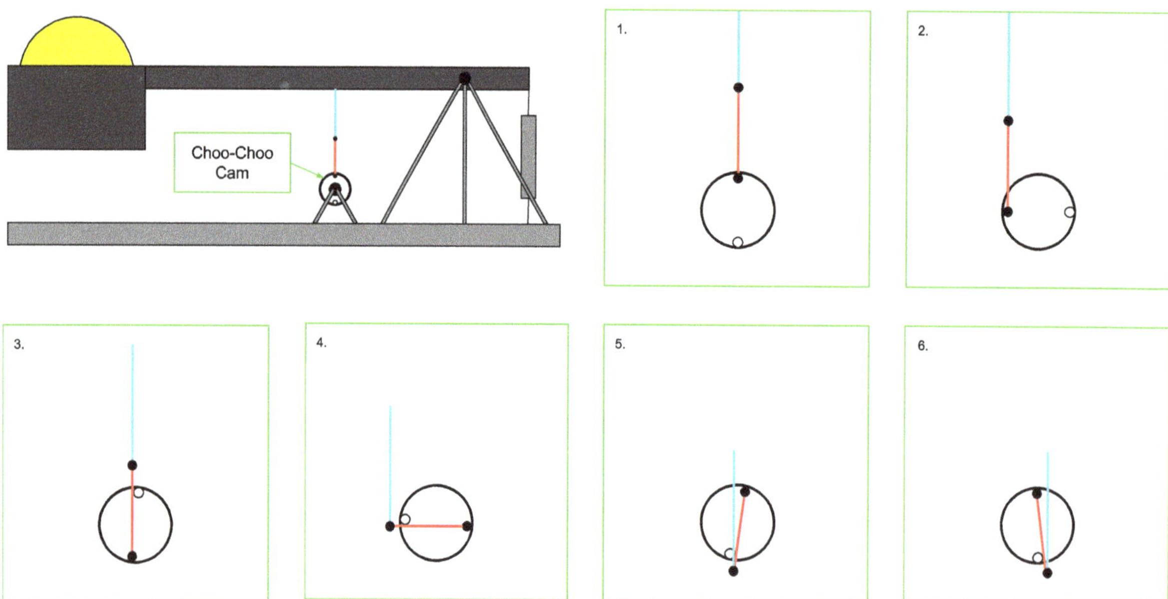


Diag. 6.3: Choo-Choo CAM Catapult Abstract

This design is the first of 2 types of cam release designs for catapults. A choo-choo cam, (name taken from its similarity to locomotive train propulsion systems), features 2 free turning pivot arms. The red pivot arm can be stopped by the white circle which represents a physical stopper. The cyan pivot arm can bypass the white stopper because it moves on a different plane. This cyan pivot arm is also connected to the catapult's structure. There are 6 key positions for the choo-choo cam depicted in diagram 6.3.

1. The fully released or starting position.
2. The choo-choo cam starts to pull back the catapult.
3. The white stopper first engages with the red pivot arm and continues to pull the catapult down.

4. The white stopper continues to engage the red pivot arm and continues to pull the catapult down.
5. The position where the catapult's potential energy is at its maximum as the choo-choo cam has brought the cyan pivot arm as far down as possible. Any further and the choo-choo cam will release the catapult. This is often the position that a catapult using a choo-choo cam will be at when loading the game object so that a quick release can be triggered.
6. The choo-choo cam turns just a bit further. Now the white stopper is no longer in the way of the red pivot arm. The catapult suddenly snaps upward releasing the stored potential energy from the spring. The choo-choo cam then returns to position 1.

## 6.3 Standard (Large Snail) Cam Catapults

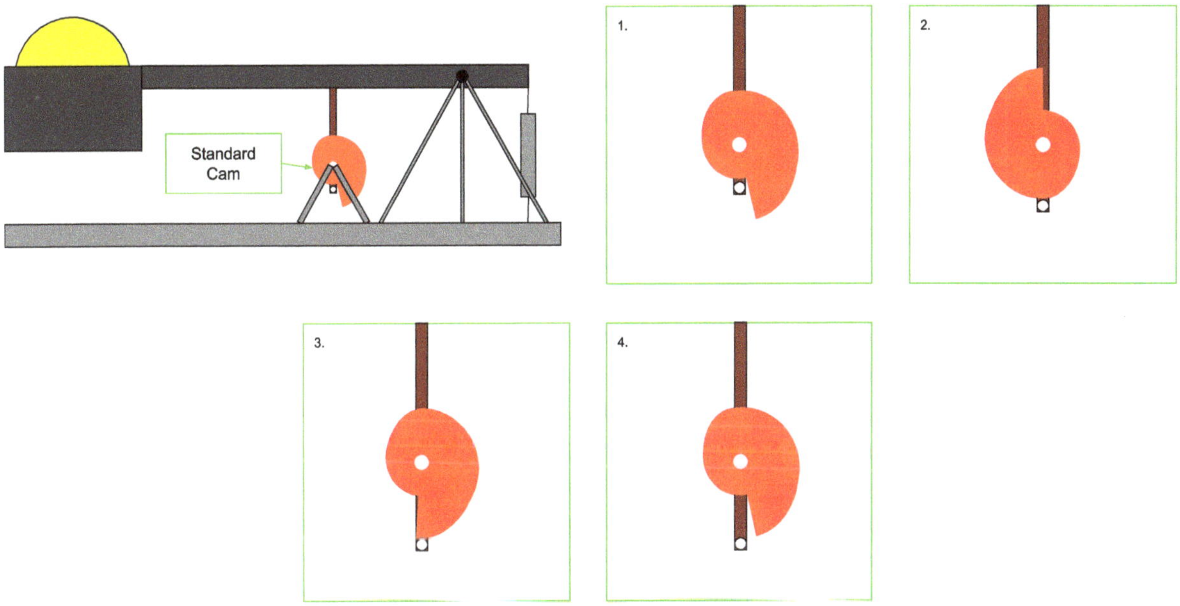


Diag. 6.4: Standard CAM Catapult Abstract

This release mechanism uses a large snail cam to drive the catapult. Typically when cam is mentioned in the FRC context, people think of the snail tensioning cam used for chain assemblies. Therefore, this cam variant is referred to as the standard cam release mechanism. The snail cam rotates on a pivot secured to the base structure. The dark red vertical rectangle represents an arm attached to the catapult structure. Protruding from the arm is a white spacer. There are 4 key positions for the snail cam depicted in diagram 6.4.

1. The cam is in the same plane as the white spacer. This can be thought of as the starting position.
2. The cam rotates anticlockwise further pulling the dark red arm downward.
3. The cam is at the loading point where the catapult's potential energy is at its maximum as the snail cam has brought the dark red arm as far down as possible. Any further and the snail cam will release the catapult. This is often the position that a catapult using a snail cam will be at when loading the game object so that a quick release can be triggered.
4. After turning the cam further, nothing is stopping the spring from pulling the catapult back up. The stored potential energy is released and the mechanism returns to position 1.

## 6.4 Sheared Gear Catapults

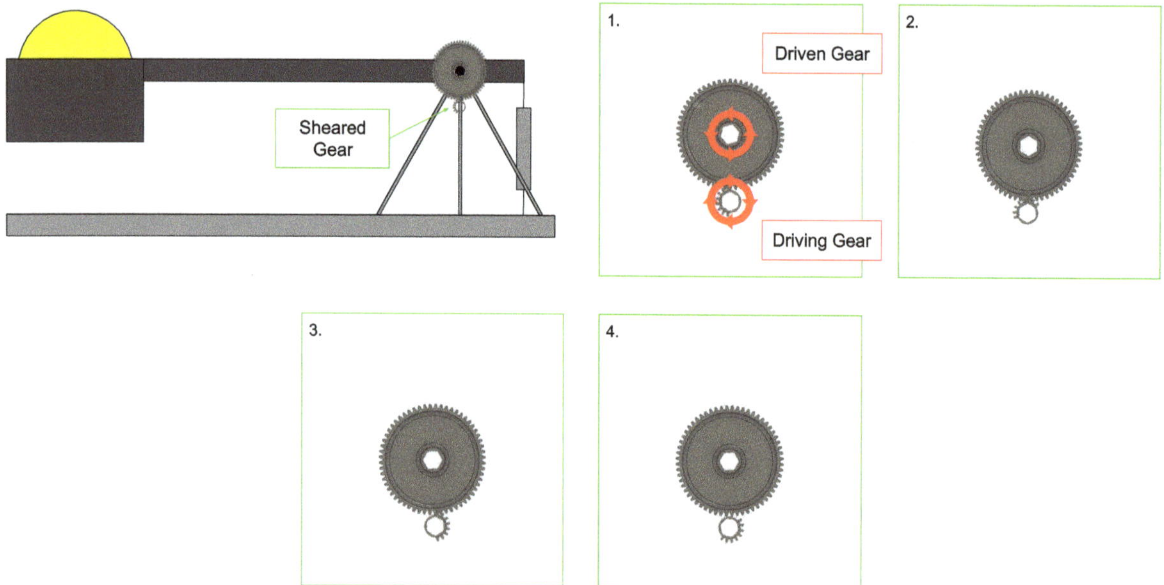


Diag. 6.5: Sheared Gear Catapult Abstract

This release mechanism uses a sheared gear to drive the catapult. In the magnified position shots, the top gear is at the catapult's pivot and turns the pivot. The bottom gear is the sheared gear that runs off of a gearbox and motor(s). A gearbox is needed to achieve the torque required to turn the catapult (not pictured in diagram 6.5). There are 4 key positions for the gears depicted in diagram 6.5.

1. This can be considered the starting position where the driving sheared gear's teeth are just starting to mesh against the top driven gear. Red circular arrows are used to show which way each gear is turning.
2. The driving gear turns a bit further clockwise. The driven gear turns anticlockwise pushing the catapult down.
3. The driving gear is turned to the loading point where the catapult's potential energy is at its maximum. The catapult has been rotated anticlockwise the furthest possible amount. If the driving gear turns any further, it will release the catapult. This is often the position that a catapult using a sheared gear release mechanism will be at when loading the game object so that a quick release can be triggered.
4. The driving gear turns a bit further, now the section of the gear that is sheared off is facing the driven gear. This makes the driven gear free spinning. The potential energy of the spring is suddenly released and pulls the catapult back into position, firing the game object. Turning the driving gear further repeats the process and returns to position 1.

Note there is a 4:1 gear ratio used in this abstract with half of the driving gear sheared off. This setup limits the catapult pivot to just ⅛ of a full rotation, as the modified gear only engages for half of its cycle. Adjusting the gear ratio and how much of the driving gear is sheared off affects how much the catapult can rotate until the release mechanism is triggered.

## 6.5 Winch Catapults

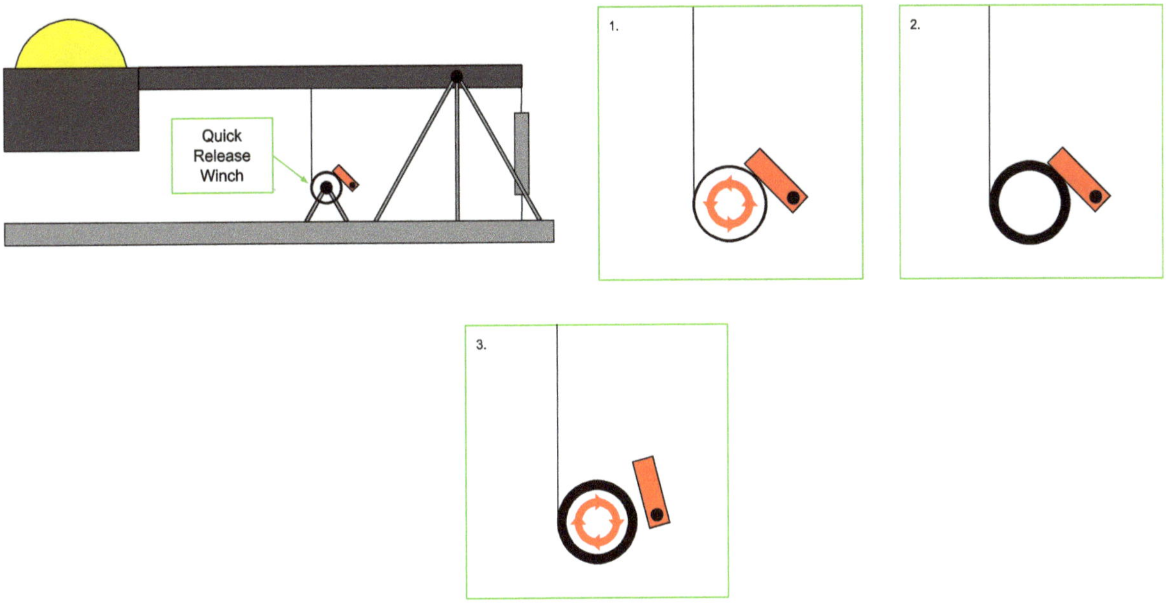


Diag. 6.6: Quick Release Winch Catapult Abstract

This release mechanism uses a winch that winds up a rope to pull the catapult down. When the catapult is ready to fire, the winch is released, letting all of the rope out suddenly. There are 3 key positions:

1. The red block represents a stopper. Various methods can be used to create the necessary stopping action such as using a motor, pneumatic, or ratchet system. (See Subchapter 11.2 for more information on ratchets). The winch is spun by a motor in the counter clockwise direction as shown by the red circular arrows.
2. At some point enough rope is coiled around the winch that the catapult is considered loaded. There is theoretically no limit to how far the winch can spin to bring the catapult down and stretch the spring. However, either some structure will get in the way first, like the catapult structure colliding with the base structure, or in the worst case, the rope in the winch snapping due to it being under too much stress from the catapult's spring. Whatever this "loaded" position might be can be tracked by using encoders on the winch's turning axle.
3. The stopper is removed and the winch mechanism is disengaged, allowing it to become free spinning. This leads to the winch spinning out its rope, releasing the potential energy stored in the spring. The mechanism then returns to position 1.

There are some benefits and complications with the quick release winch design. The main benefit is that shots can be varied with this type of design given the winch can pull the rope in by different amounts, storing varied amounts of potential energy in the catapult's springs before release. The complications lie in achieving a quick release. One way lies in having a gear that drives the winch when spun one direction and disengages with the winch when spun the other direction by use of ratchets. (See Subchapter 11.2 on Ratchets.) This vaguely follows the idea in diagram 6.6. Another way is to use pneumatics and dog shifting gears to engage the winch when needed and to achieve the quick release by disengaging the dog shifting gears (see Subchapter 11.4 on Gear Shifting).

## 6.6 Pneumatic Catapults

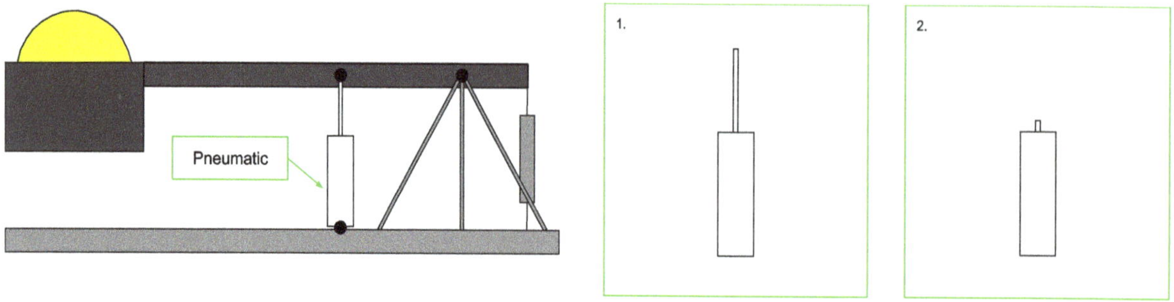


Diag. 6.7: Pneumatic Catapult Abstract

The pneumatic release mechanism in concept is pretty simple. A pneumatic pulls in the catapult for loading and releases the catapult for firing. Both ends of the pneumatic are on pivots as shown by the black dots. There are only 2 key positions.

1. The pneumatic is extended or released. This is the starting or firing position.
2. The pneumatic is retracted, stretching the spring of the catapult and maximizing the potential energy stored. This is often the position that a pneumatic release catapult will be at when loading in the game object so that a quick release can then be triggered. Extending the pneumatic leads back to position 1.

The complication lies in the execution of a pneumatic release catapult. A pneumatic release catapult is all about airflow. A relatively large and strong pneumatic or multiple pneumatics would be needed to overcome the force of the spring. The air needs to be able to escape from the pneumatic cylinder at a rate equal to or faster than that of the spring pulling the catapult back up. Otherwise the pneumatic will not facilitate a quick release of the stored potential energy. Making sure pneumatics do not leak and that proper airflow is maintained is a tedious task for Electronics & Programming. It is considered a somewhat difficult task.

The air pressure in a pneumatic system can also be varied by the pneumatic control module, which theoretically could be used to vary the tensioning of the catapult's spring to some degree. In reality, this is not very reliable and the variation in tension is limited.

## 6.7 Catapult Example

With all the theoretical catapult mechanisms now explained, here is a fully fledged design example by Team 118. Refer to https://www.frc-handbook.com/ref/23 for more information on the robot and game.

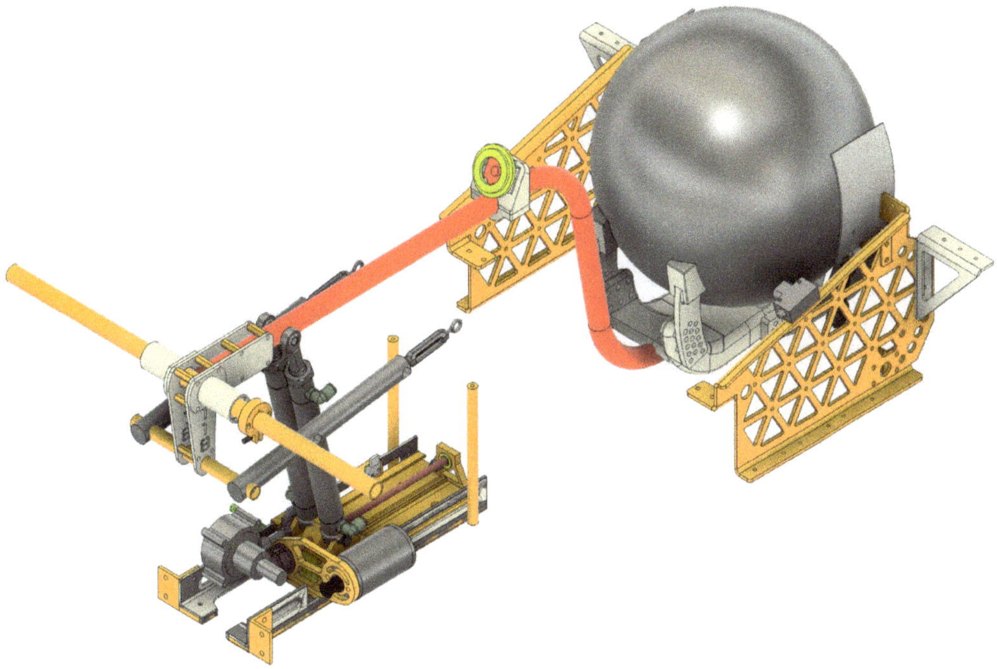

Diag. 6.8: Catapult Ex. 1 Isometric View; Ref. 23

This is a special pneumatic release catapult that is able to adjust its shot trajectory.

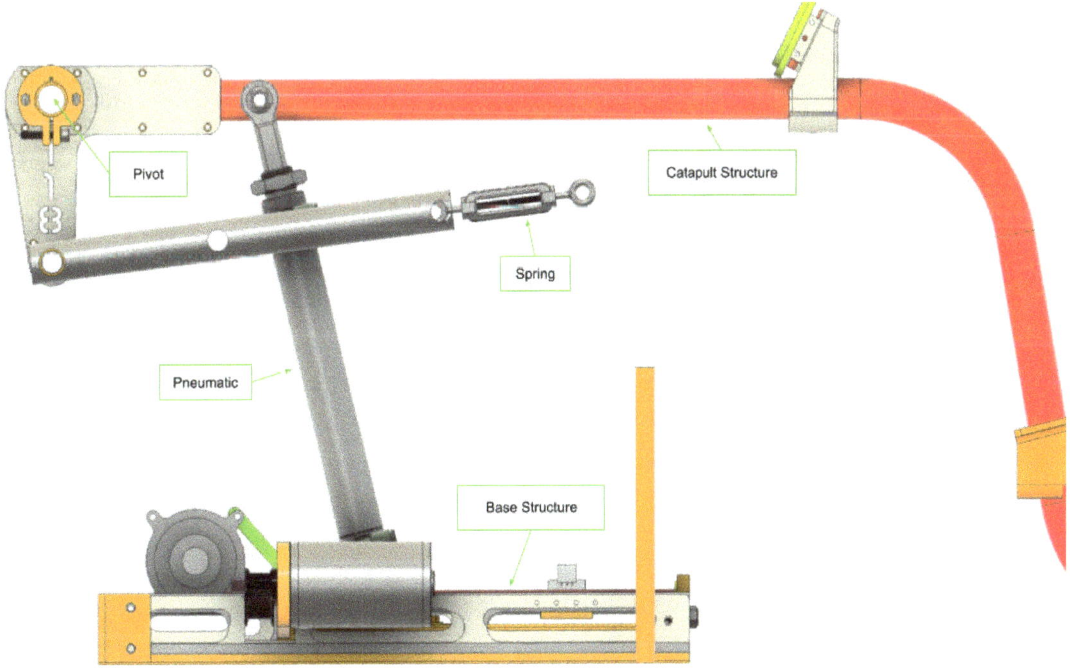


Diag. 6.9: Catapult Ex. 1 Side View; Ref. 23

This catapult has all the standard components of a pneumatic release catapult. Compared against diagram 6.2, this view of the catapult is almost a direct mirror of the abstract diagram. The catapult mechanism sits on the base structure. The pivot is in the top left of the catapult design. A spring provides the source of potential energy in the catapult and drives the pivot anticlockwise when under load. The pneumatics act as the release mechanism for this catapult.

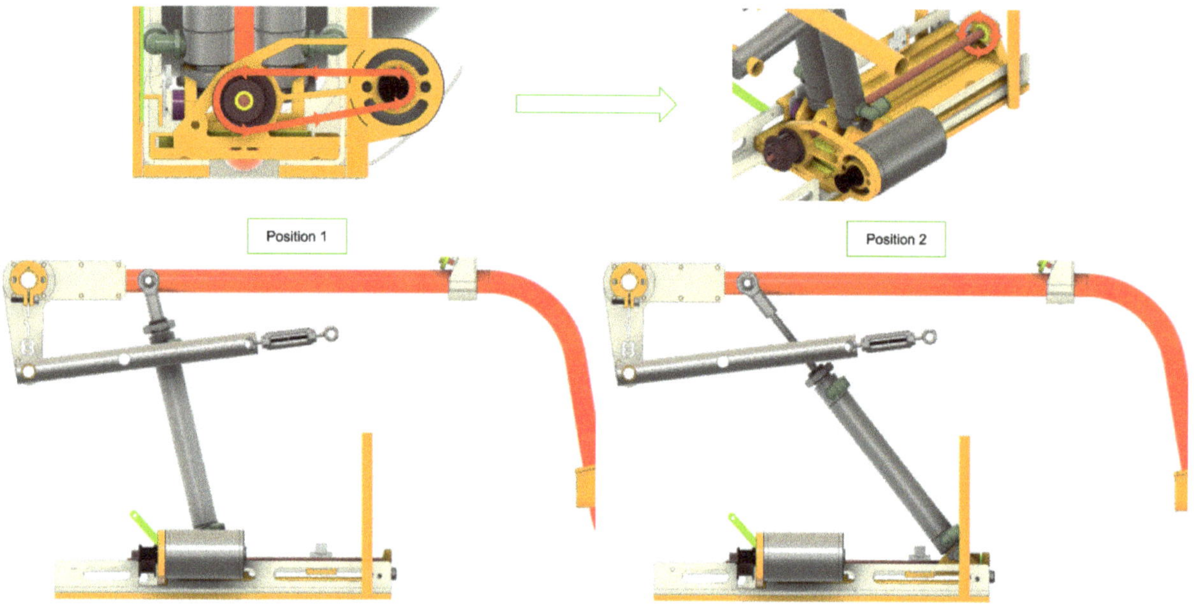


Diag. 6.10: Catapult Ex. 1 Collage View; Ref. 23

With superb precision engineering, this catapult is able to adjust its shot trajectory by adjusting the position of the bottom pivot of the pneumatics. The bottom pivot of the pneumatics is on a slide driven by a screw. This screw is turned by a motor. These elements are shown in the top 2 images of diagram 6.10. This results in the two extreme positions of the pneumatic shown in the bottom 2 images of diagram 6.10.

## 6.8 Turntables for Catapults

In theory it is possible to mount a catapult on a turntable. However, in practice, this can be quite awkward.

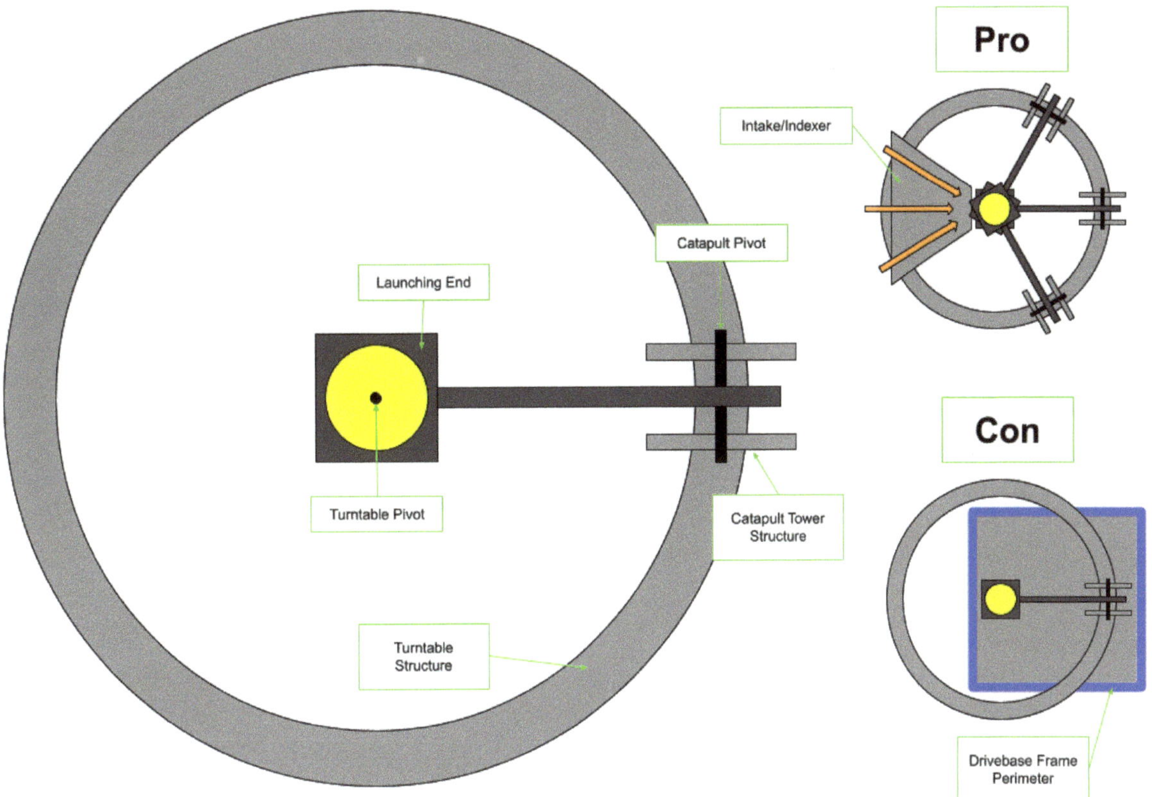


Diag. 6.11: Catapult Turntable Abstract 1

If the pivot point for the turntable is placed at the launching end of the catapult, balls can be fed to the catapult no matter its orientation as can be seen in the Pro section of diagram 6.11. However, the turntable must now be very large as the pivot point of the turntable is usually a sizable distance away from the catapult's tower, (referred to as the catapult's pivot supports in diagram 6.1). Such a large turntable may not even fit within the robot's frame perimeter as can be seen in the Con section of diagram 6.11.

Technically, this drawback can be worked around by designing a short enough catapult such that the large turntable fits entirely inside the drivebase frame perimeter, but that will limit the effectiveness of the catapult. A short-arm catapult launches game objects with lower velocity, thereby limiting its range and trajectory. This new issue can be resolved by boosting the catapult's maximum potential energy through the addition of extra springs. Unfortunately, this introduces further engineering challenges such as ensuring the catapult's structure is durable enough to withstand the increased forces and preventing the springs from damaging the catapult during launches.

Another way to work around the drawback of the large turntable is to instead use a smaller turntable with a beam extending off the edge of the turntable to reach the catapult's tower, but that solution would be pretty unstable.

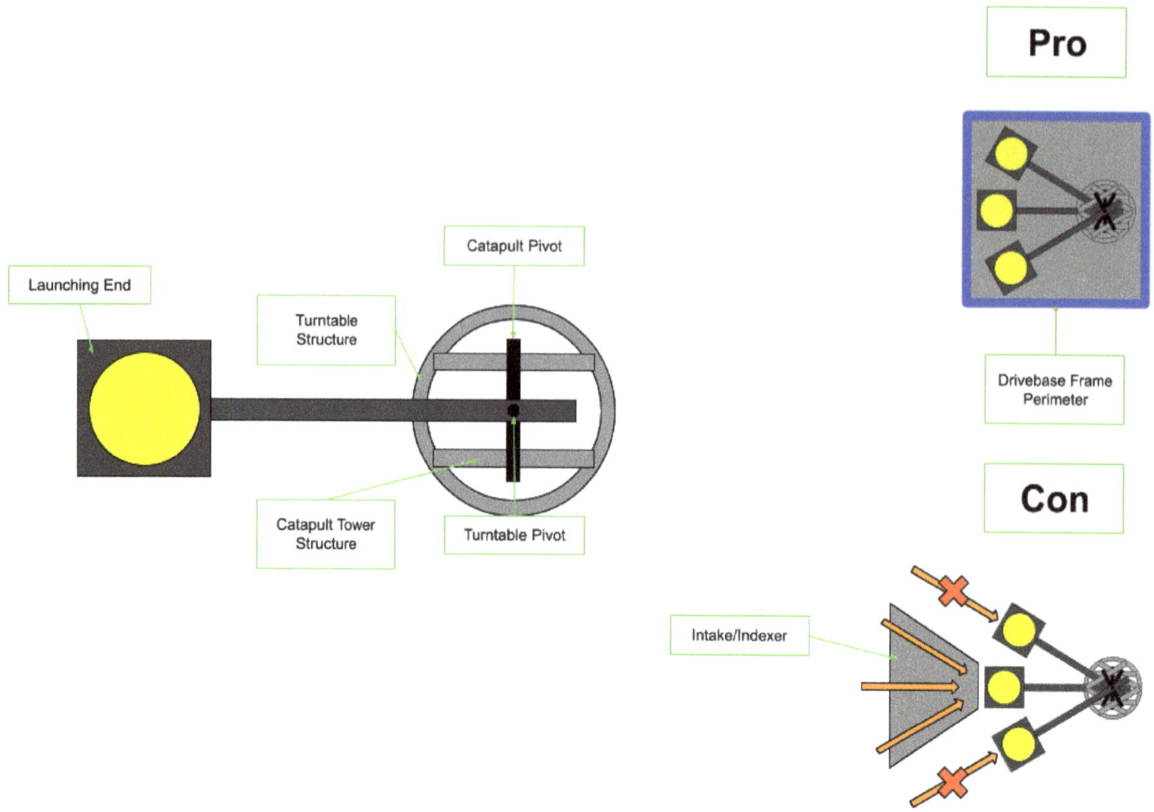


Diag. 6.12: Catapult Turntable Abstract 2

A completely different approach to integrating a catapult with a turntable is to place the pivot point of the turntable directly under the catapult's pivot, at a more stable section of the assembly. The problem with this solution is that the catapult will not always align with the intake, meaning the catapult will always have to realign with the intake/indexer to accept a game object before turning back to align for a shot. This may not be a big hindrance if the robot is only allowed to hold one game object at a time. Where game objects are required to be fired in rapid succession, this set up will not work.

Both options are problematic, resulting in either a complicated design or a functionally limited robot.

## 6.9 Summary on Catapults

Below is a summary of the various catapults discussed. Note the table reflects what is generally true for all mechanisms of said type, however additional innovation is always possible to further improve the capabilities of a mechanism. For example, Reference 23 describes shifting the pneumatic pivot to enable improved variable catapult shots.

<table>
<tr><th colspan="3">Catapults</th></tr>
<tr><td colspan="3"><ul><li>Stores up potential energy before release</li><li>Used for heavy game objects</li><li>Trajectory can be varied if tension on the spring at the point of release can be varied</li></ul></td></tr>
<tr><td rowspan="2">Cam Release Catapult</td><td>Choo-Choo Cam Release Catapult</td><td><ul><li>Utilises 2 free turning pivot arms</li><li>Arms are pulled down by rotating the cam to stretch spring</li><li>2nd arm moving past stopper triggers release</li></ul></td></tr>
<tr><td>Standard/Snail Cam Release Catapult</td><td><ul><li>Utilises snail cam tensioning structure</li><li>Snail cam rotated to stretch spring</li><li>Rotating past the Snail cam's peak triggers release</li></ul></td></tr>
<tr><td>Gear Release Catapult</td><td colspan="2"><ul><li>Simplest to design</li><li>Utilises a large driven gear and a small sheared driver gear</li><li>Rotating the gears stretches the spring</li><li>When the sheared side of the small gear faces the large gear triggers release</li></ul></td></tr>
<tr><td>Winch Release Catapult</td><td colspan="2"><ul><li>Utilizes a winch and stopper</li><li>Winds up winch to stretch the spring</li><li>Allows for varied tensioning of spring by how much the winch is wound up</li><li>Stopper is moved off the winch triggers release</li></ul></td></tr>
<tr><td>Pneumatic Release Catapult</td><td colspan="2"><ul><li>Utilizes pneumatics</li><li>Retracts pneumatics to stretch spring</li><li>Theoretically allows for varied tensioning of spring by control air pressure; in practice unreliable</li><li>Extends pneumatics to trigger release</li><li>Regulating pneumatic airflow and preventing leaks is difficult</li></ul></td></tr>
<tr><th colspan="3">Turntables</th></tr>
<tr><td colspan="3"><ul><li>Supplementary mechanism to orientate catapults</li><li>Rarely implemented, as a suitable turntable is either larger than size limits, or triggers stability/alignment issues when kept within size limits</li></ul></td></tr>
</table>

Diag. 6.13: Catapult Flowchart

On the next page is a flowchart that provides a decision tree on the choice of catapult. Note that priorities differ depending on application, so strictly following this flowchart in every instance may not be ideal. In general, catapults should only be considered when the game involves the launching of heavy game objects, otherwise, flywheels are typically the superior choice.

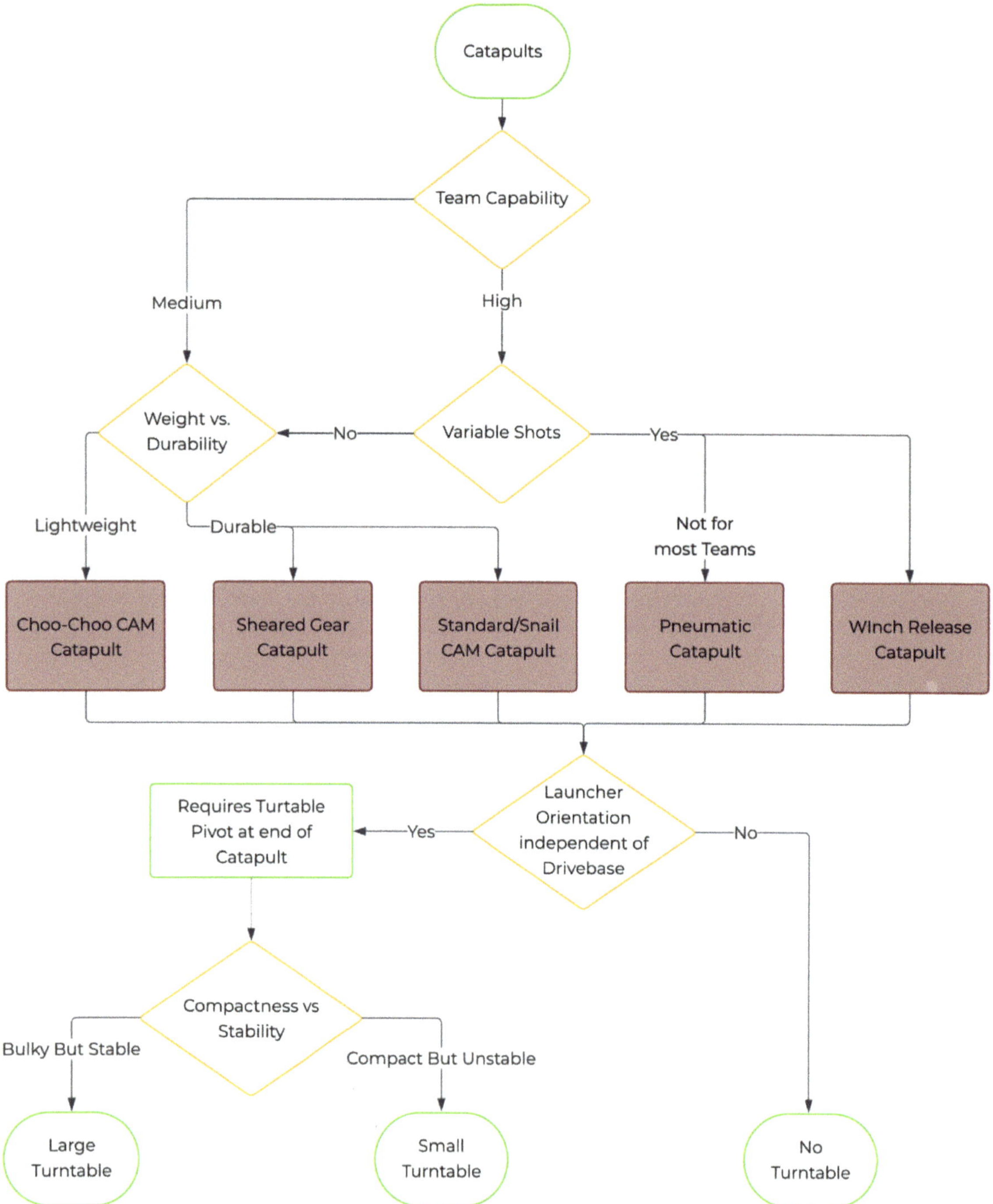


Diag. 6.14: Catapult Flowchart

# 7. Arms

## 7.1 Introduction

Arms are mechanisms that involve moving bar(s) around pivot(s). Almost always, arms are used to move some other mechanism in a robot, whether that's a claw, or an overbumper intake. The mechanism being moved is referred to as the end effector.

Most arms include a gearbox. Gear reduction is required since FRC motors spin very fast and do not have a lot of torque. The longer the arm, the greater the required torque to move the arm. Similarly, the heavier the load at the end of the arm, the greater the required torque to move it. Higher required torque translates to a higher gear ratio. An alternative to using motors and gearboxes to power arms is to use pneumatics. Pneumatics are suitable only when an arm needs to move between 2 fixed positions, since pneumatics can only be either fully retracted or extended. A motor and a gearbox can rotate the pivot to be at any angle, leading to greater flexibility with how the arm is used.

This chapter is not about the design of a gearbox that powers an arm, but instead it is about the different types of arms that could potentially be used in an FRC game.

## 7.2 Single Bar Arms/Pivot Arms

A single bar arm, also known as a pivot arm, is the simplest of all arm types. It is literally just a single bar turning around a pivot.

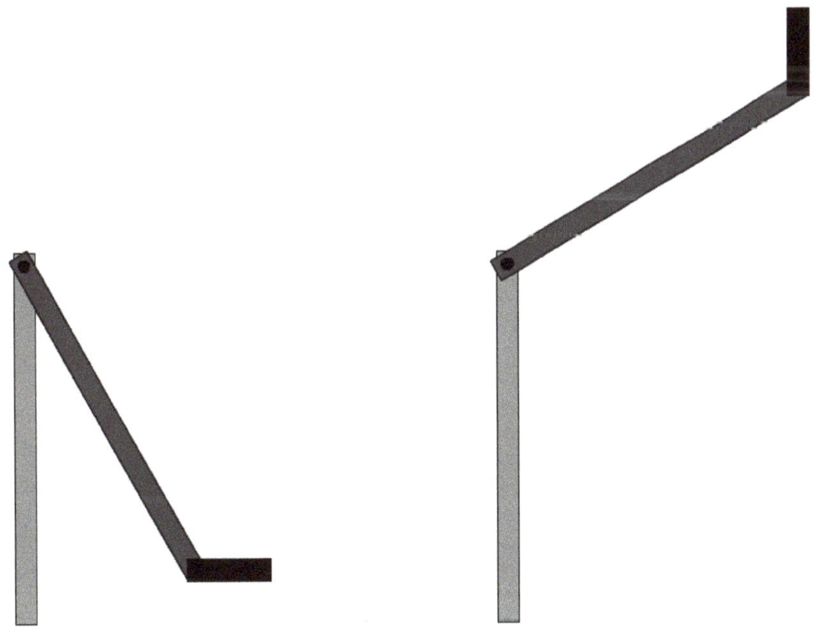

Diag. 7.1: Pivot Arm Abstract

An extra black bar is drawn at the end of the arm to show that an attached mechanism would rotate with the same orientation of the arm.

Most simple overbumper intakes make use of a pivot arm since the only important orientation of the pivot arm is when the intake is flipped down for intaking. Furthermore, the pivot arm is the easiest arm to make.

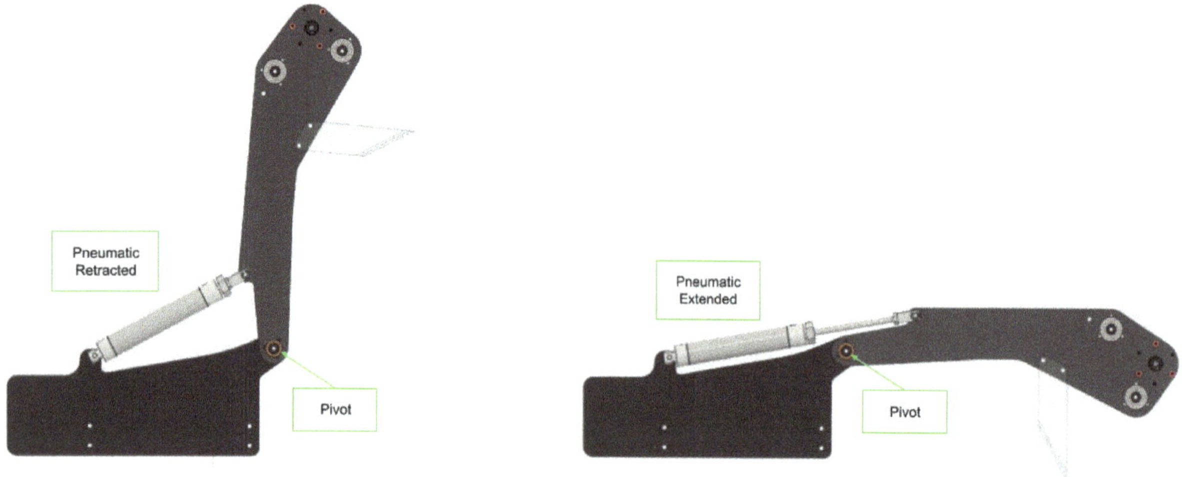


Diag. 7.2: Pivot Arm Ex. 1 Side View; Ref. 18

Above is an example of a pivot arm for an overbumper intake that utilizes pneumatics. As only 2 positions are required for an overbumper intake, (starting and intaking positions), using pneumatics to drive this pivot is appropriate.

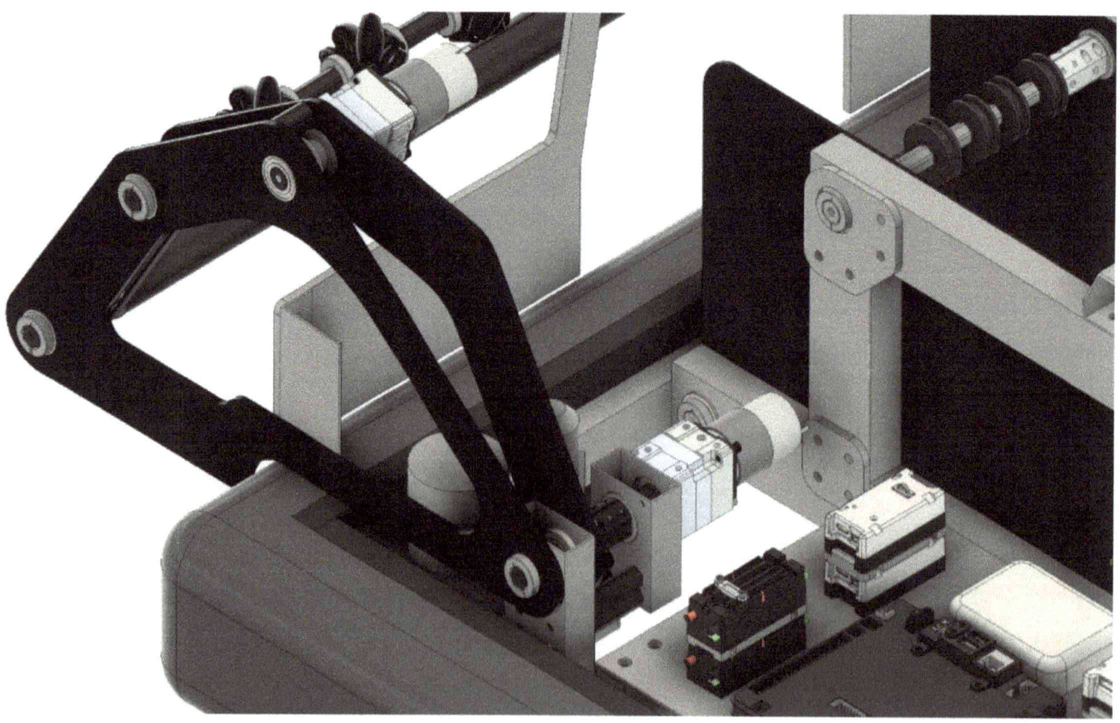

Diag. 7.3: Pivot Arm Ex. 2 Isometric View; Ref. 17

On the other hand, using a gearbox also works for driving a pivot arm.

Diag. 7.4: Pivot Arm Ex. 2 Side View; Ref. 17

The gearbox turns this pivot so that the intake can achieve the 2 key positions (starting and intaking positions). However, as this pivot arm uses a gearbox it can turn to any position between the 2 shown positions or even further clockwise than the right position shown (until the intake hits the indexer).

## 7.3 4-Bar Arms; Chain Bars

While pivot arms are effective for pivoting and are very simple mechanisms, there is a potential problem – anything stuck on the arm turns with the arm. This can be a problem if the end effector is a claw as most claws want to maintain the same orientation even when they are brought up to a higher position. A claw stuck on the end of a pivot arm would not be able to achieve this. The rest of the arms covered in this chapter solve this problem with a simple concept.

### 4-Bar Arm Mechanism

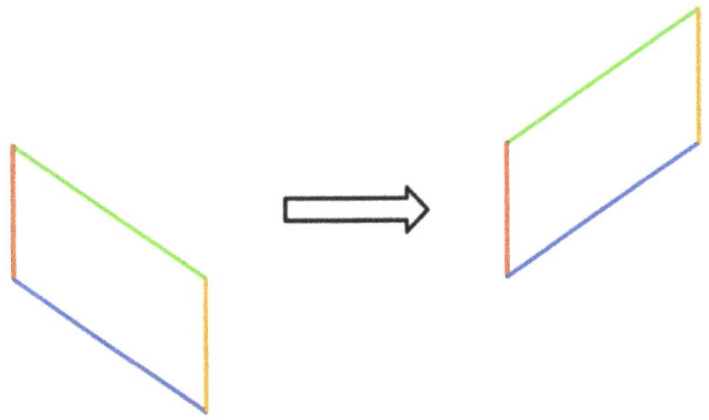

Diag. 7.5: Parallelogram Abstract

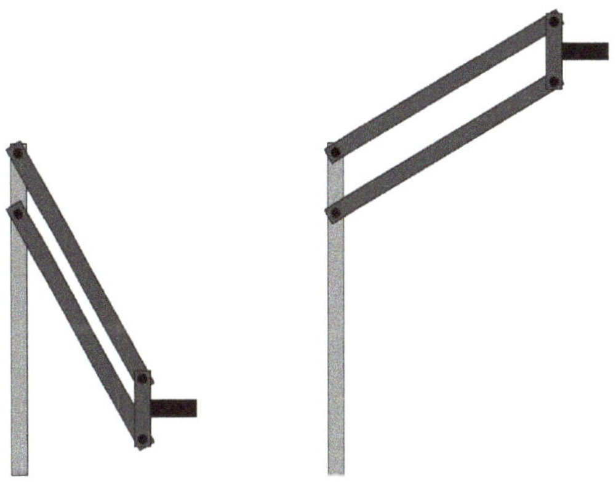

Diag. 7.6: 4-Bar Arm Abstract

The key shape to introduce is the parallelogram. By its nature its opposite sides are always parallel to each other. Referring to diagram 7.5, if one side of the parallelogram is frozen in place, (the red line), while one of the adjacent edges is rotated, (the blue line), the opposite side from the frozen side would still have the same orientation, (the orange line). This is useful because in a sense, (assuming all corners are freely rotating pivots), as long as one of the sides is frozen in place, whatever way the parallelogram is manipulated the opposing side keeps the same orientation.

Now applying that concept to a practical mechanism, (refer to diagram 7.6), we get the 4-bar arm. There are 4 pivots, shown as black dots, and a fixed side of the parallelogram, represented by the lighter gray bar. This configuration keeps the opposite side at the same orientation no matter the arm's rotation. The black bar parallel to the ground in the image shows this property of the 4-bar arm.

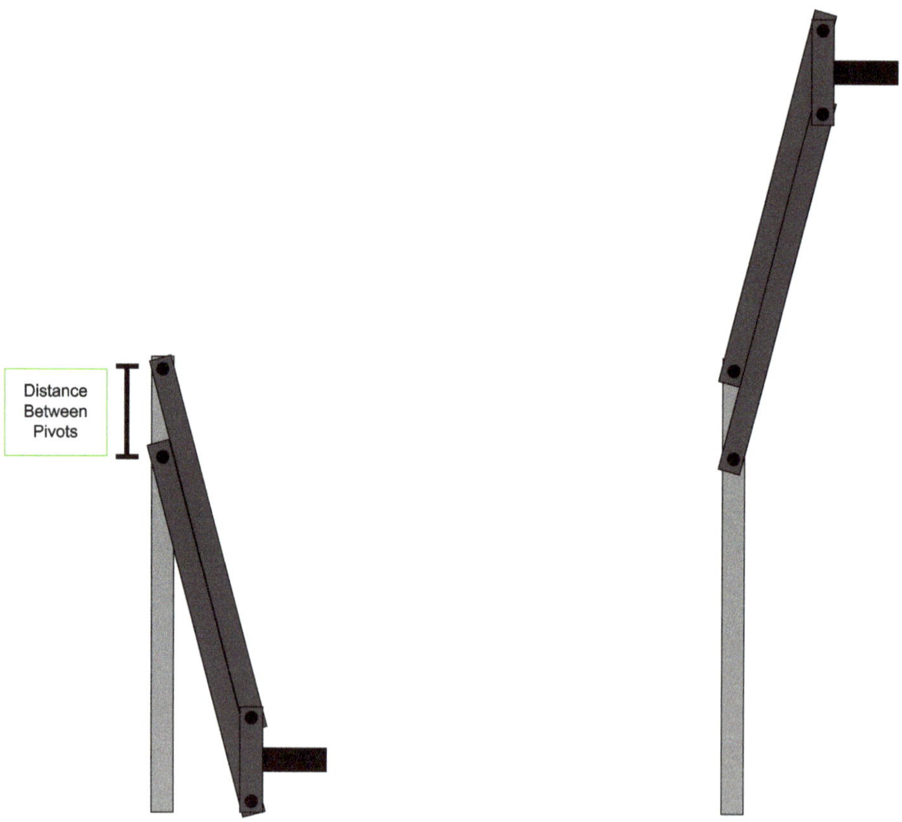


Diag. 7.7: 4 Bar Arm Abstract 2

Assuming the 2 rotating arms of the 4-bar arm are in the same plane, there is a limit as to how high and how low the arm can rotate based on the distance between the two pivots. At the upper and lower limits shown, the 2 arms collide, stopping the arm from going any further.

There are other ways to use a parallelogram shape to achieve an even greater range of motion. One way is to place the 2 rotating arms in different planes by using axles of different lengths. An even simpler way is to use a chain bar.

### Chain Bar Arm Mechanism

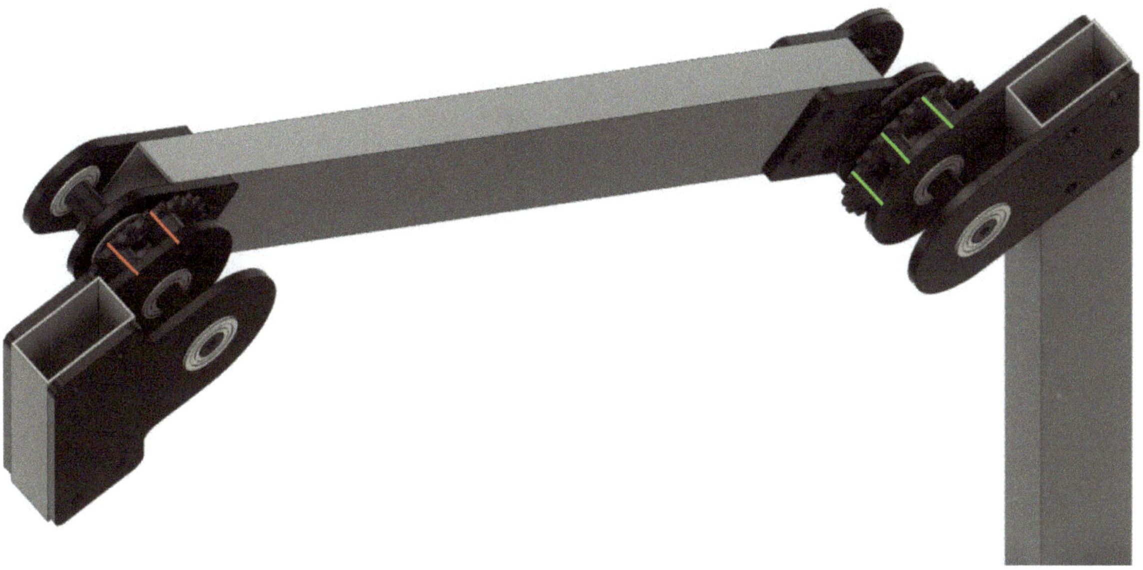

Diag. 7.8: Chain Bar Ex. Top Slanted View Ref. 24

The chain bar works by linking 2 sprockets with a continuous chain, with one mounted on the tower and the other on the end effector. As the gearbox rotates the arm, the chain forces the end effector to pivot in sync, keeping it oriented in the same direction throughout motion. The green lines highlight the standoffs used to mount the right sprocket to the tower, and the red lines highlight the standoffs used to mount the left sprocket to the end effector. (There are a total of 6 standoffs on each side. Not all are pictured.) The sprockets' bore must be big enough for the axle to go through so the sprockets do not turn with the axles. The arm itself is bound to the axle by a hex bore adapter, assuming Thunderhex shafts are used. A gearbox can then power the axle to rotate the chain bar like a pivot arm.

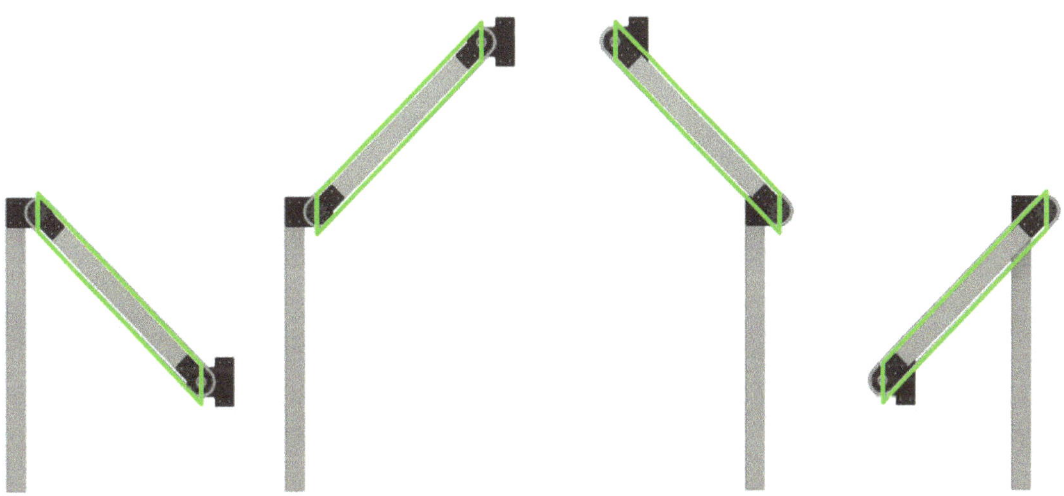

Diag. 7.9: Chain Bar Ex. Side View Multiple Cuts; Ref. 24

The chain that wraps around the 2 mounted sprockets makes it so that the end effector will be kept in the same orientation no matter where the arm is turned to. One way to think about how this works is that the chain forces the end effector to turn in the opposite direction that the arm itself turns, keeping the end effector parallel to the tower. Another way to think about it is shown in diagram 7.9. If a parallelogram is drawn with the chain, it can be thought of as a 4-bar arm with the difference being that a chain will never collide with itself.

Diag. 7.10: Chain Bar Ex. Movement; Ref. 24

The range of motion that can be achieved is shown by diagram 7.10. To see an animated version, refer to https://www.frc-handbook.com/vol1/7/d10..

The chain bar arm achieves a wider range of movement as compared to the standard 4-bar arm. Generally, as a chain bar uses fewer parts and is lighter, it is almost always preferred over the 4-bar arm when a simple linkage arm is required.

**Uneven 4-Bar Arm Mechanism**

In a parallel 4-bar arm, the end effector stays on the same plane throughout the arm's movement. However, a 4-bar arm can be designed so that the transition between positions results in a rotation of the end effector. Diagram 7.11 showcases such an example.

Diag. 7.11: Uneven 4 Bar Ex. Side View; Ref. 10

The green lines are of unequal lengths and show the orientation of the 2 arms while the red line shows the relative orientation of the end effector. The advantage of such a system is that it can make the robot more compact. In this example, a rather large overbumper intake can be brought into the robot and reorientated so that it fits within frame perimeter. An uneven 4-bar arm requires dimensioning a system in multiple positions.

## 7.4 6-Bar Arms

A 6-bar arm has a few trade offs compared to the 4-bar arm.

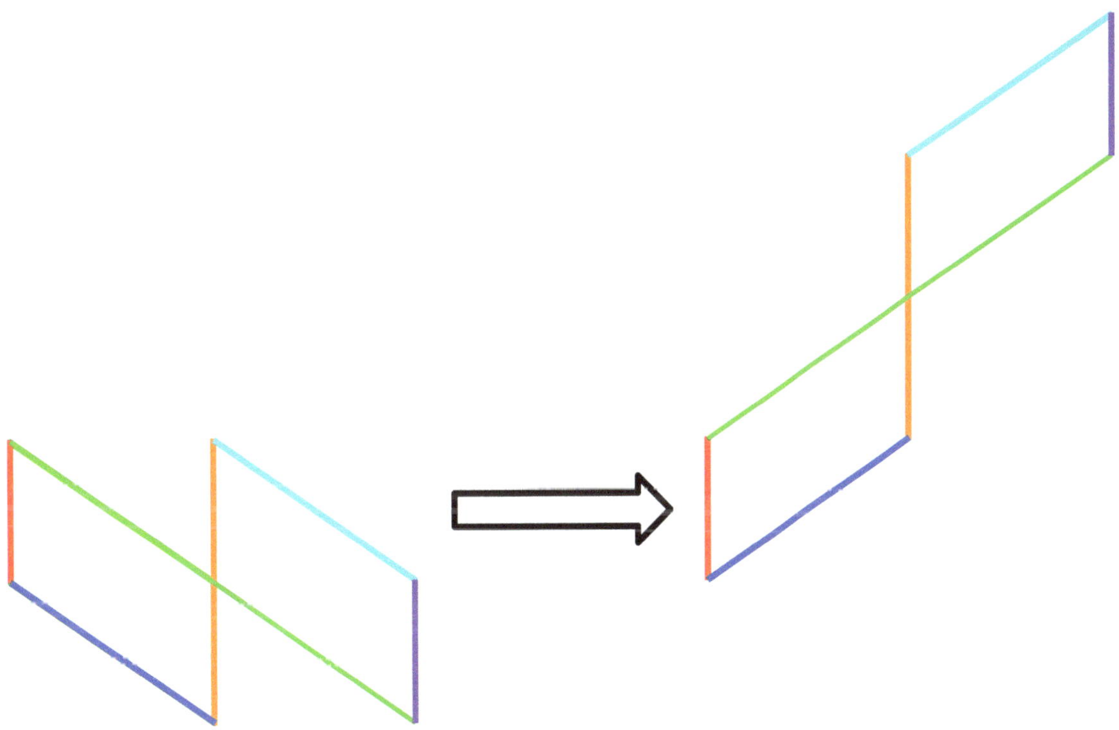

Diag. 7.12: 2 Merged Parallelograms Abstract

The 6-bar arm takes the idea of merging 2 parallelograms together to form a shape that is made up of 6 bars. Compared to the 4-bar version, (diagram 7.5), the orange and green lines in diagram 7.12 have been doubled in length. 2 additional bars are added to complete the second parallelogram. Again, the red line is treated as the stationary bar.

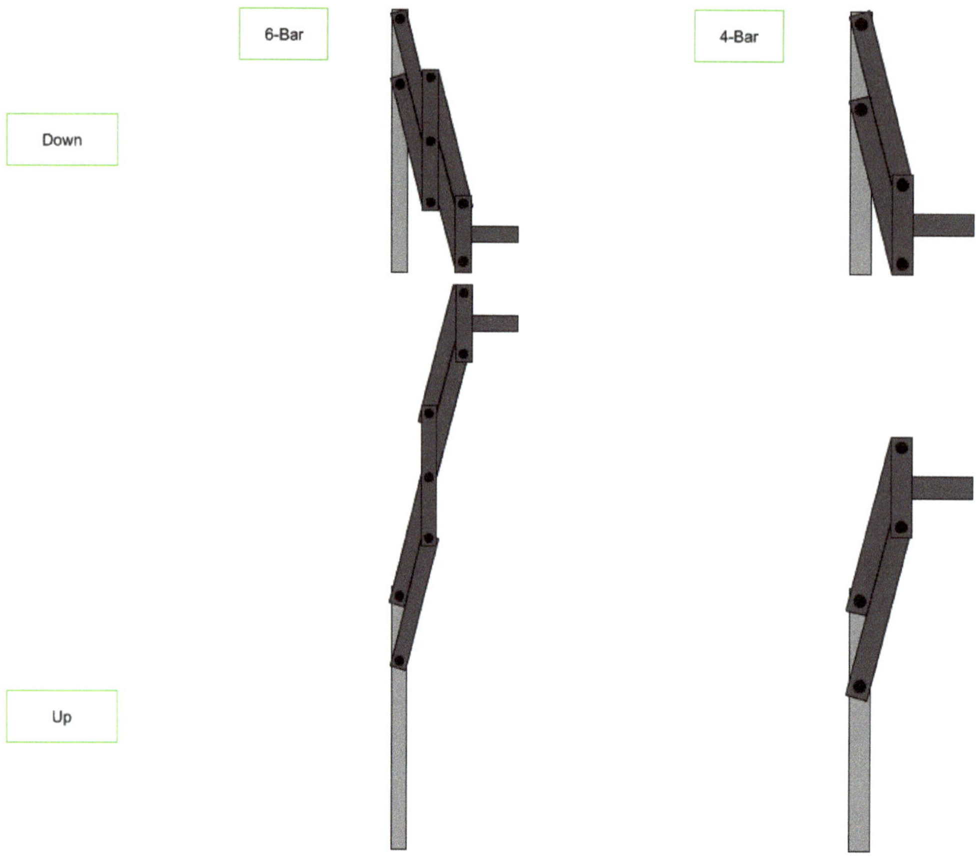


Diag. 7.13: 6-Bar vs. 4-Bar Arm Abstract

The only time to use a 6-bar arm instead of a 4-bar arm is when there is some kind of space constraint and the extra reach of a 6-bar arm is required. In the down positions both the 6-bar arm and 4-bar arm have approximately the same height and width. However, in their maximum up positions the 6-bar arm reaches a bit higher.

The drawback of using a 6-bar arm is the additional complexity with the extra parts used. If the extra maximum height is not needed, the additional complexity is not warranted in a robot. Usually a 4-bar arm can accomplish whatever is required in an FRC game. There are also other ways to give a 4-bar arm additional reach (see Chapter 9 on Multi-Axis Arms).

## 7.5 Double Reverse 4-Bar (DR4B) Arms

The double reverse 4-bar (DR4B) arm uses two 4-bar linkages that rotate in reverse sync—meaning they move in opposite directions to generate vertical extension. While this configuration adds mechanical complexity, it allows for significantly greater reach compared to a standard 4-bar arm.

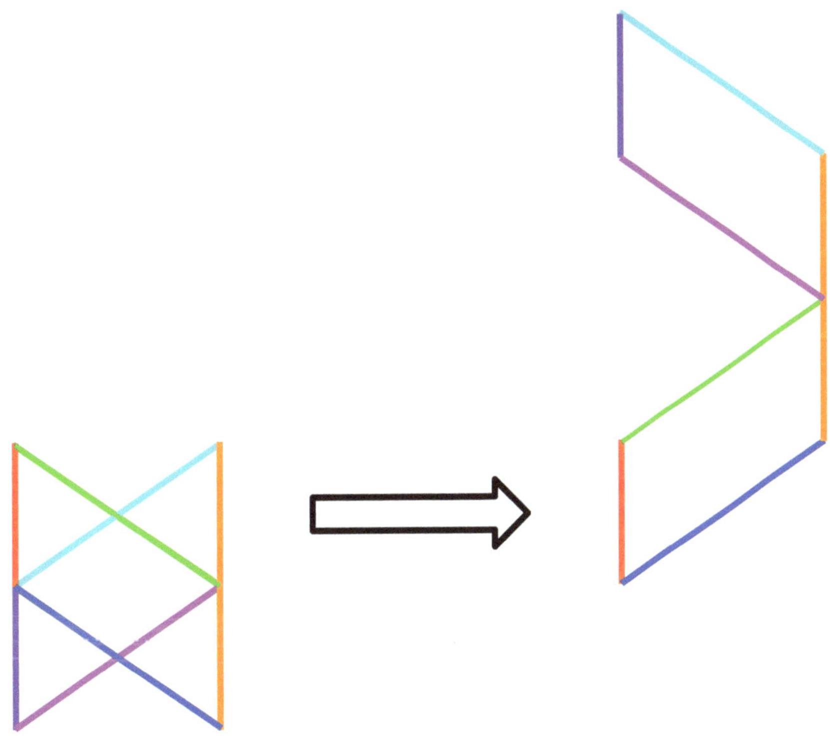

Diag. 7.14: 2 Merged Parallelograms Alternate Abstract

The DR4B takes the idea of merging 2 parallelograms together to a new level. In this setup, the green and purple lines turn in sync with each other. In sync in this case means that the green line and pink line constantly maintain the same interior angle with the shared orange line. This leads to the 2 positions shown in diagram 7.14.

## Using 4 Bar Linkages for DR4B Arm

The synchronization of the 2 separate 4-bar arms is achieved by using gears mounted at the interaction point between the two 4-bar arms.

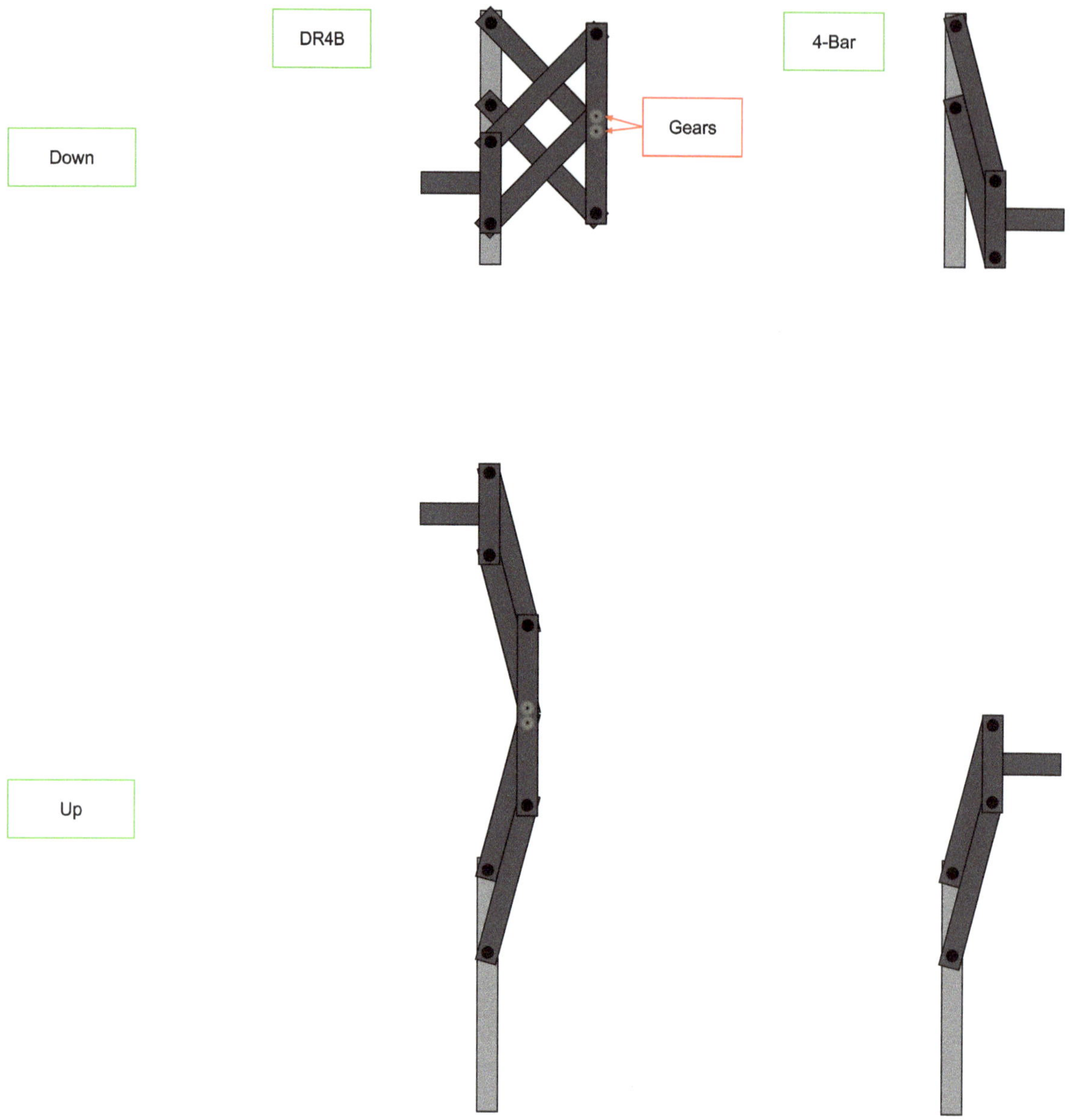


Diag. 7.15: DR4B vs. 4-Bar Abstract

The gears are mounted such that they turn with the individual arms. When the bottom 4-bar rotates, the top 4-bar is forced to rotate too. The light brown gears show the position of these gears in the DR4B design.

As shown in comparison to a regular 4-bar arm, the starting down position may result in a slightly wider width. (It is assumed that the tower height across both the DR4B arm and the 4-bar arm is the same. In actuality, the fully collapsed tower height for a DR4B would be taller at its narrowest width with all 4 arms touching each other in the down position.) In the up position, the DR4B arm effectively doubles the reach of a 4-bar arm.

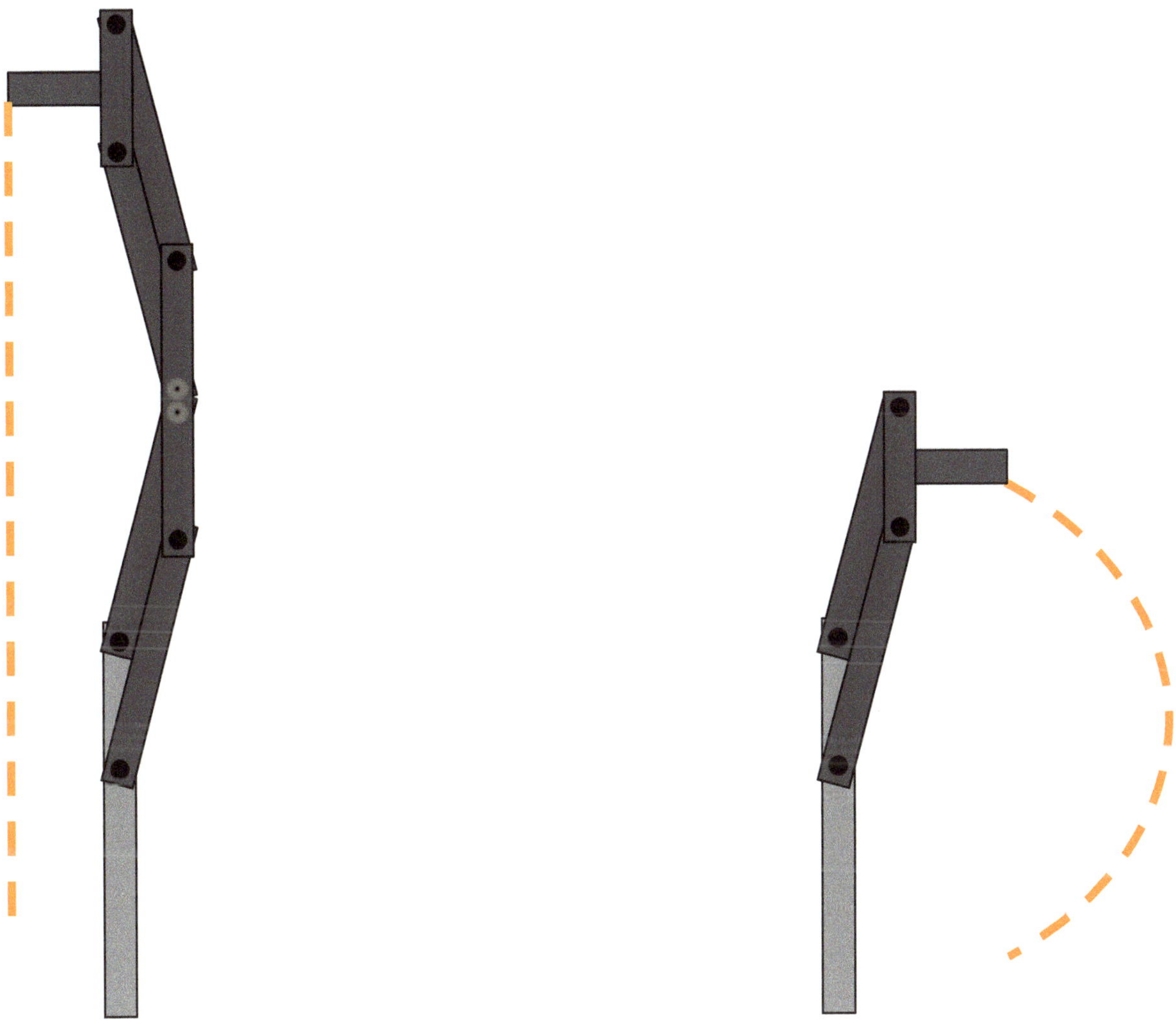

Diag. 7.16: DR4B vs. 4-Bar End Effector Movement Abstract

A special feature to note is that the end effector of a DR4B arm will move perfectly straight up if the 2 sets of 4-bars are of equal length. In comparison a 4-bar's end effector moves in an arc.

It should be noted that a DR4B can be made with a chain bar mechanism instead of a 4-bar mechanism.

## Using Chain Bars for DR4B Arm

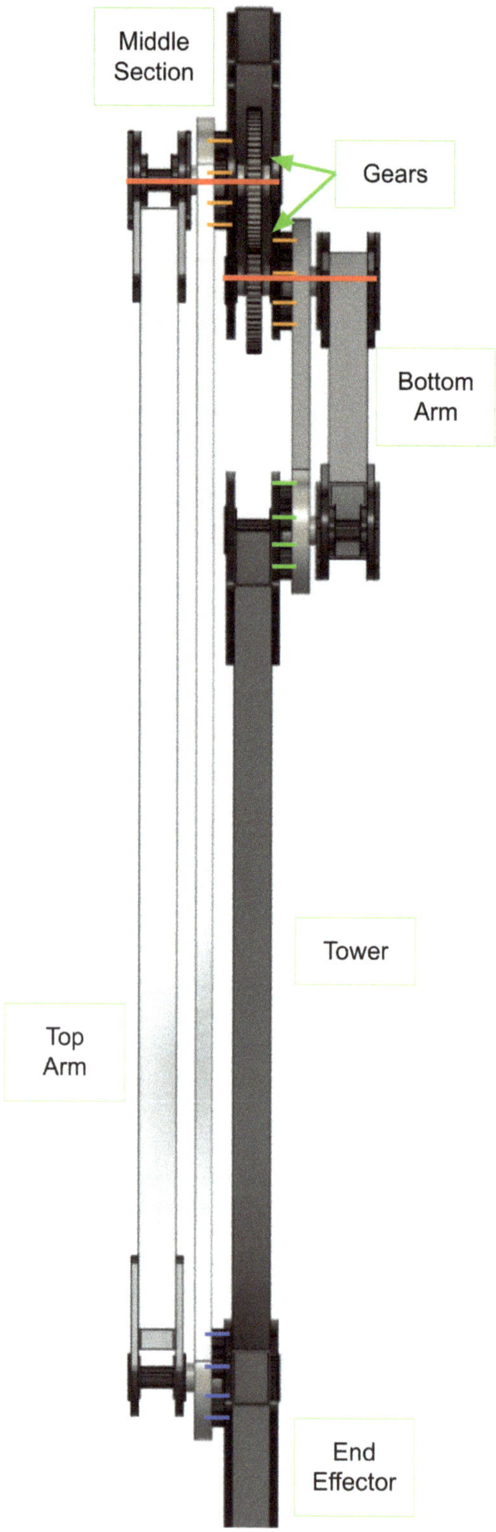


Diagram 7.17 shows how the sprockets for the chain bars are mounted. The green lines highlight the standoffs connecting a sprocket to the tower, the orange lines highlight the standoffs connecting sprockets to the middle section of the DR4B arm and the blue lines highlight the standoffs connecting the sprocket to the end effector. As with any chain bar, the sprockets must be kept static to the tower/middle section/end effector and not rotate with the arm.

The red lines highlight the axles going through the end of the bottom arm and the start of the top arm. The axles are locked to the rotation of the arm by VersaHubs. This enables the gears positioned between the plates of the middle section to drive the 2 sets of chain bars in sync.

Diag. 7.17: Chain Bar DR4B Ex. Top Slanted View; Ref. 25

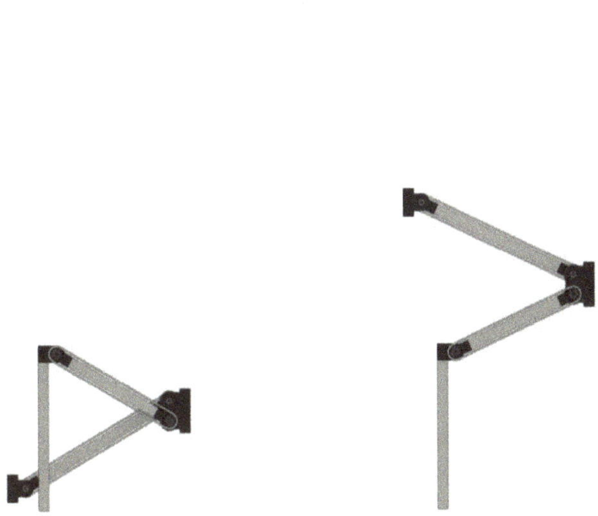

Diag. 7.18: Chain Bar DR4B Ex. Movement; Ref. 25

The range of motion that can be achieved by this DR4B arm is shown by diagram 7.18. Refer to https://www.frc-handbook.com/vol1/7/d18 to see an animated version.

Note that as the top chain bar is slightly longer than the bottom chain bar, (this was done so that the end effector does not hit the tower during the range of movement). Due to this, the end effector of this DR4B arm will move in a slight arc and not perfectly straight up.

#### Distinction Between DR4B Arms and Multi-Jointed Arms

A DR4B design is not to be confused with a multi-jointed arm, a type of multi-axis arm (see Subchapter 9.4). A multi-jointed arm allows each arm to be controlled individually. A DR4B is comparatively simpler as the entire arm runs off of a single gearbox driving the bottom set of 4-bars. The highlighted gears are what force the top set of 4-bars to move in sync with the bottom set. Conversely, each set of 4-bars cannot be individually controlled, unlike a multi jointed arm.

## 7.6 Theoretical Extensions of Arm Types

Just as how the idea of a 4-bar arm can be extended into making a 6-bar arm, a 6-bar arm could be extended similarly into an 8-bar arm and so on. Also, just as how a 4-bar arm can be stacked to make a double reverse 4-bar arm, this idea could be extended to make a triple reverse 4-bar arm. Rarely will any of these types of arms be a good idea. As a general rule, additional parts and complexity result in unforeseen problems during the build. Simpler is generally better. Among other potential problems, additional joints tend to increase the friction in a robot—something that should be minimized as much as possible.

If it is possible to solve an engineering problem with a simpler arm, that should be the first solution. That being said, a more complicated arm may at times be required for a robot to perform optimally for an FRC game. If additional complexity is going to be added, then it should be added in a way that truly elevates the performance of the arm while minimizing the added complexity. Generally when simpler 4-bar arms are not able to solve an engineering problem, teams turn to multi-axis arms (described in Chapter 9) instead of the theoretical examples described here.

## 7.7 Summary on Arms

Below is a summary of the various arms covered in this chapter.

<table>
<tr><th colspan="2">Single Bar Arm/Pivot Arm</th></tr>
<tr><td colspan="2"><ul><li>Simple design</li><li>End effector rotates around a fixed pivot point</li><li>Moves in an arc</li><li>With motors, allows continuous rotation or precise positioning</li><li>With pneumatics, limited to 2 positions only</li></ul></td></tr>
<tr><th colspan="2">4-Bar Arm</th></tr>
<tr><td>4-Bar Arm</td><td><ul><li>Simple design</li><li>Uses 4 pieces of stock to create a parallelogram</li><li>Moves in an arc</li><li>End effector remains parallel to the ground</li><li>Typically driven by a motor to access all positions</li></ul></td></tr>
<tr><td>Chain Bar Arm</td><td><ul><li>Simple design (even simpler than 4-Bar Arm)</li><li>Use 2 sprockets and a chain to form parallelogram</li><li>Moves in an arc</li><li>End effector remains parallel to the ground</li><li>Typically driven by a motor to access all positions</li></ul></td></tr>
<tr><td>Uneven 4-Bar Arm</td><td><ul><li>Complex design</li><li>2 arms of different length connect to the same end effector</li><li>Moves in an arc</li><li>End effector's rotation affected by the difference in length between the 2 arms</li><li>Typically driven by pneumatics to switch between 2 positions</li><li>Possible to use a motor to access all positions</li></ul></td></tr>
<tr><th colspan="2">6-Bar Arm, 8-Bar Arm, 10-Bar Arm etc.</th></tr>
<tr><td colspan="2"><ul><li>Extension of 4-Bar Arm design that creates additional parallelogram shaped structures</li><li>Complex design</li><li>More compact folding of the arm while allowing a higher reach</li><li>Moves in an arc</li><li>Not recommended beyond 6-bar arms</li></ul></td></tr>
</table>

| Double Reverse 4-Bar (DR4B), (TR4B) etc. |
|---|
| • Extension of 4-Bar Arm design that creates additional 4-Bar Arms stacked on top of the previous arm facing the opposite direction linked by gears<br>• Complex design<br>• More compact folding of arms while allowing higher reach<br>• Moves vertically if the sets of arms are of equal length and there is an even number of sets<br>• Not recommended beyond DR4B |

Diag. 7.19: Arm Summary Table

The following flowchart illustrates one possible approach to selecting an arm type for a robot design. However, it does not encompass all methods of end-effector manipulation, which can also be achieved using lifts (see Chapter 8) or multi-axis arms (see Chapter 9).

Note that "reach" in the flowchart refers to the maximum height of the arm when extended. Robot size limitations predetermine the length of the arm bars, and hence the reach of an extended arm.

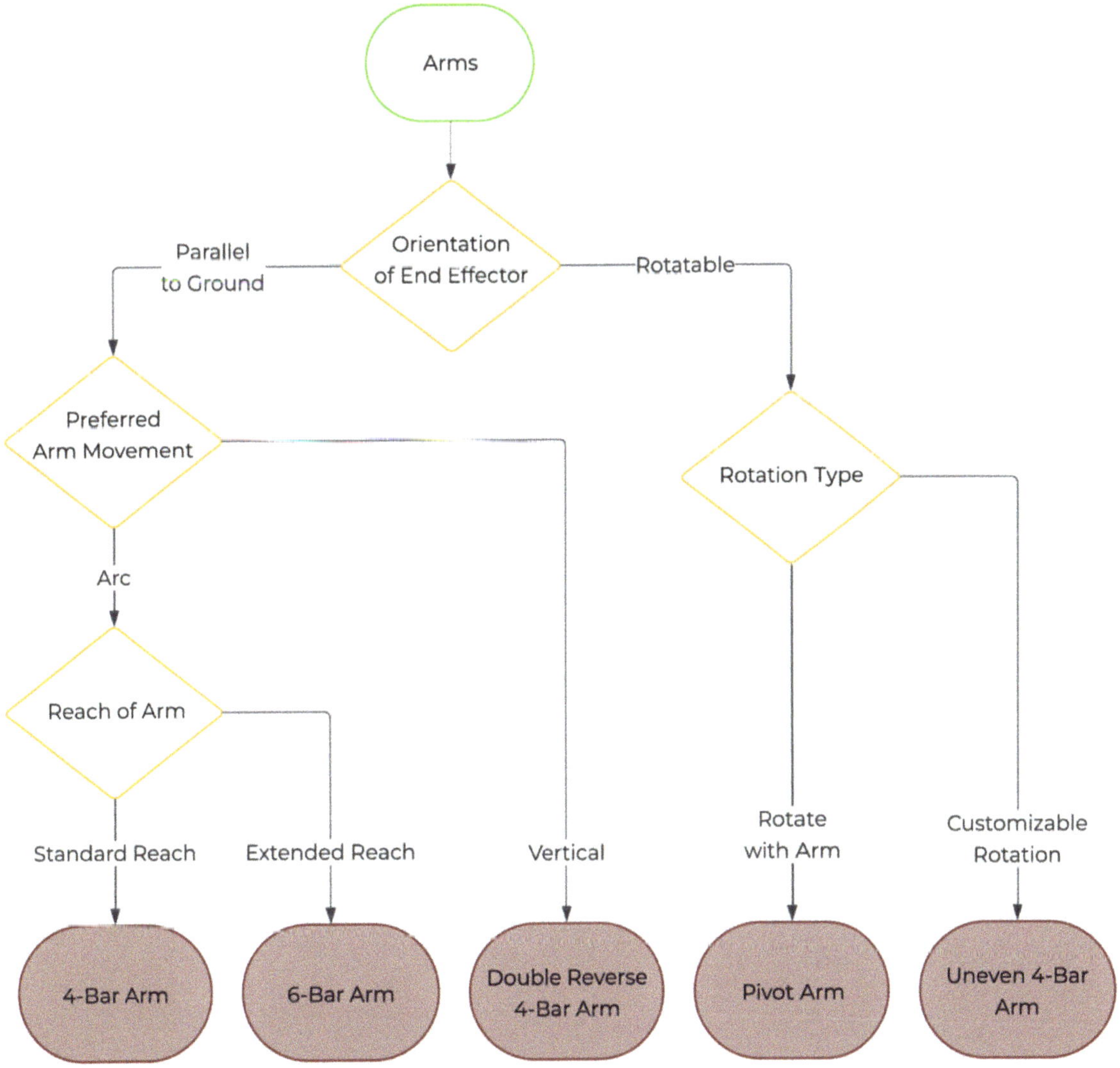


Diag. 7.20 Simple Arm Flowchart

# 8. Lifts/Elevators

## 8.1 Introduction

Lifts or Elevators are mechanisms that involve moving a structure or carriage linearly. (The word lift or elevator can be used interchangeably for all of the mechanisms in this chapter.) Almost always, lifts are used to move some other mechanism in a robot, typically a claw. The mechanism being moved is referred to as the end effector.

Most lifts are fitted with a gearbox to increase torque via gear reduction. This is required since FRC motors spin very fast and do not have a lot of torque. This chapter is not about the design of a gearbox that powers a lift, but instead about the different types of lifts that could potentially be used in an FRC game.

While there are some alternatives to creating smooth rolling action between different parts of a lift, the standard method is to use bearing modules.

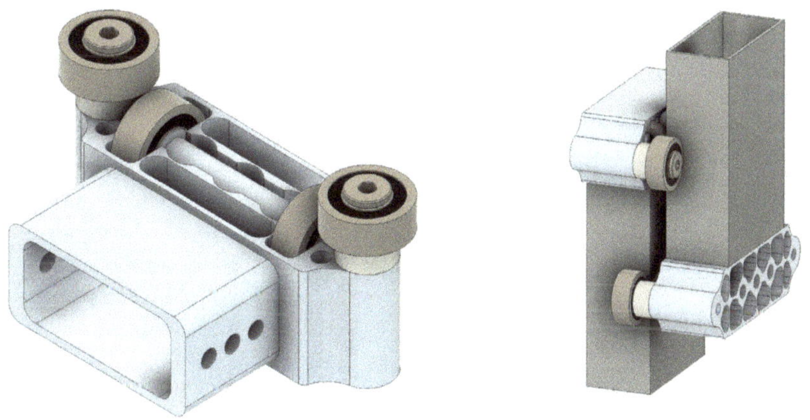

Diag. 8.1: Bearing Module Ex. 1 Multi View; Ref. 26

The above bearing module is from WestCoast Products.

A slightly cheaper and more flexible alternative is to use custom-designed bearing modules. This involves using either 3D printed parts or laser cut plates to create a structure that accurately positions the bearings in the module to function effectively.

A lift consists of 1 or more stages. In FRC terminology, the stage count refers only to the number of moving section(s) of the lift. The tower or static stage is not included when counting stages.

All lifts also make use of some kind of driving system. This is usually a rope, but can also be a chain or a belt.

## 8.2 Single Stage Lifts

The single stage lift is the most simple of all lift types. There are a few different ways to structure a single stage lift.

### Single Chain Single Stage Lift

The first method is to structure a single stage lift by using a single chain to pull the single stage upward and downward.

Diag. 8.2: Single Stage Lift Ex. 1 Isometric View; Ref. 12

This example makes use of a single stage lift as a part of a climb mechanism. See https://www.frc-handbook.com/ref/12 for more information on the game and full robot.

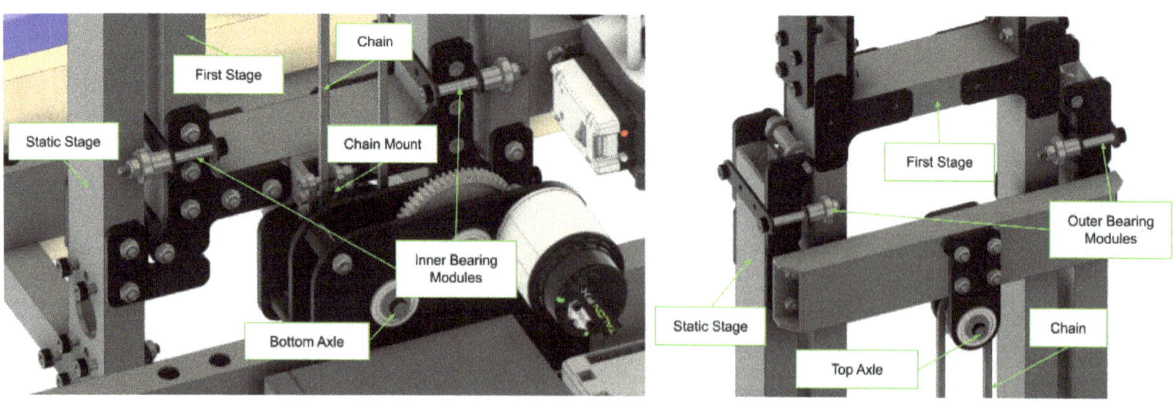


Diag. 8.3: Single Stage Lift Ex. 1 Zoomed In View; Ref. 12

Diagram 8.3 shows a zoomed in view of the top and bottom of the lift with labels of the key parts. The inner and outer bearing modules create the smooth linear motion between the static and first stage. The chain wraps around the top and bottom axles which have to be mounted on structures that are stationary relative to the static stage (or tower). The chain mount piece connects the first stage to the chain. That way when the chain is moved, the first stage follows along with the chain.

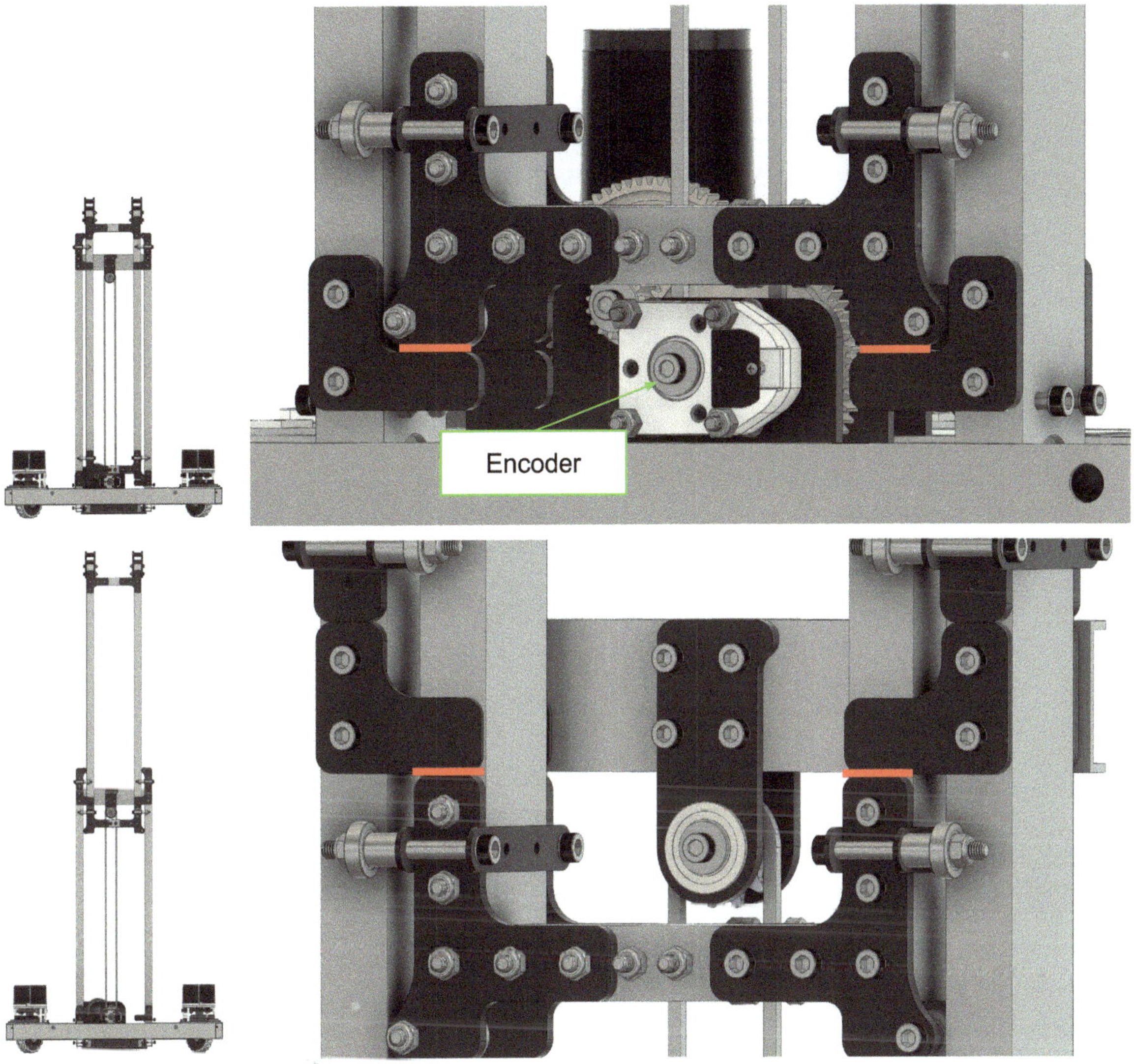


Diag. 8.4: Single Stage Lift Ex. 1 Multi View; Ref. 12

Diagram 8.4 shows the single stage lift at its 2 extremes. The red lines highlight the physical limiters that stop the lift from going any further. While it is possible to have a lift's motion controlled by limiters in programming, it is almost always better to have physical limiters as well.

The encoder tracks the position of the single stage lift.

While this is an example of a single chain single stage lift, it is possible to switch the driving system to use pulleys and belts instead of sprockets and chain. Mounting the first stage to a pulley is not as easy as mounting to a chain (as a chain has holes to put bolts through), but it is a possible alternative.

### Double Spool Single Stage Lift

The second method is by using 2 spools of strings to control the movement of the lift. (Note, another term for spool is winch.)

Diag. 8.5: Single Stage Lift Ex. 2 Isometric View; Ref. 11

This example makes use of a single stage lift as a multi-tool that both functions as this robot's climbing mechanism and supports the color wheel mechanism. (For more information on the game and full robot see https://www.frc-handbook.com/ref/11)

Note, this example is actually a type of multi-axis arm design as described in Subchapter 9.2, (a type of extendable arm). This section describes only the lift, not the entire mechanism.

Zooming in to the mechanism in diagram 8.6 on the right, it can be seen that the lift is extended and retracted by 2 spools of strings. The red line represents the path of the extending spool. It starts tied to the left winch, goes through the static stage's stock, up around the outer bearing module, as seen in the zoom in view on the top left, down between the static stage and first stage stock, and tied around the inner bearing module's bolt. The orange line represents the path of the retracting spool, which starts wound up with its end tied around the right winch, which is in turn mounted to the inner bearing module's bolt. The retracting spool and extending spool are geared, (with ropes properly fed), so that when one spool is being reeled in, the other is let out. When the extending spool is reeled in, it pulls the first stage upward, while the retracting spool is let out. When the retracting spool is reeled in, meaning the spools spin in reverse, the first stage is pulled downward, while the extending spool is let out. This controls the movement of the first stage.

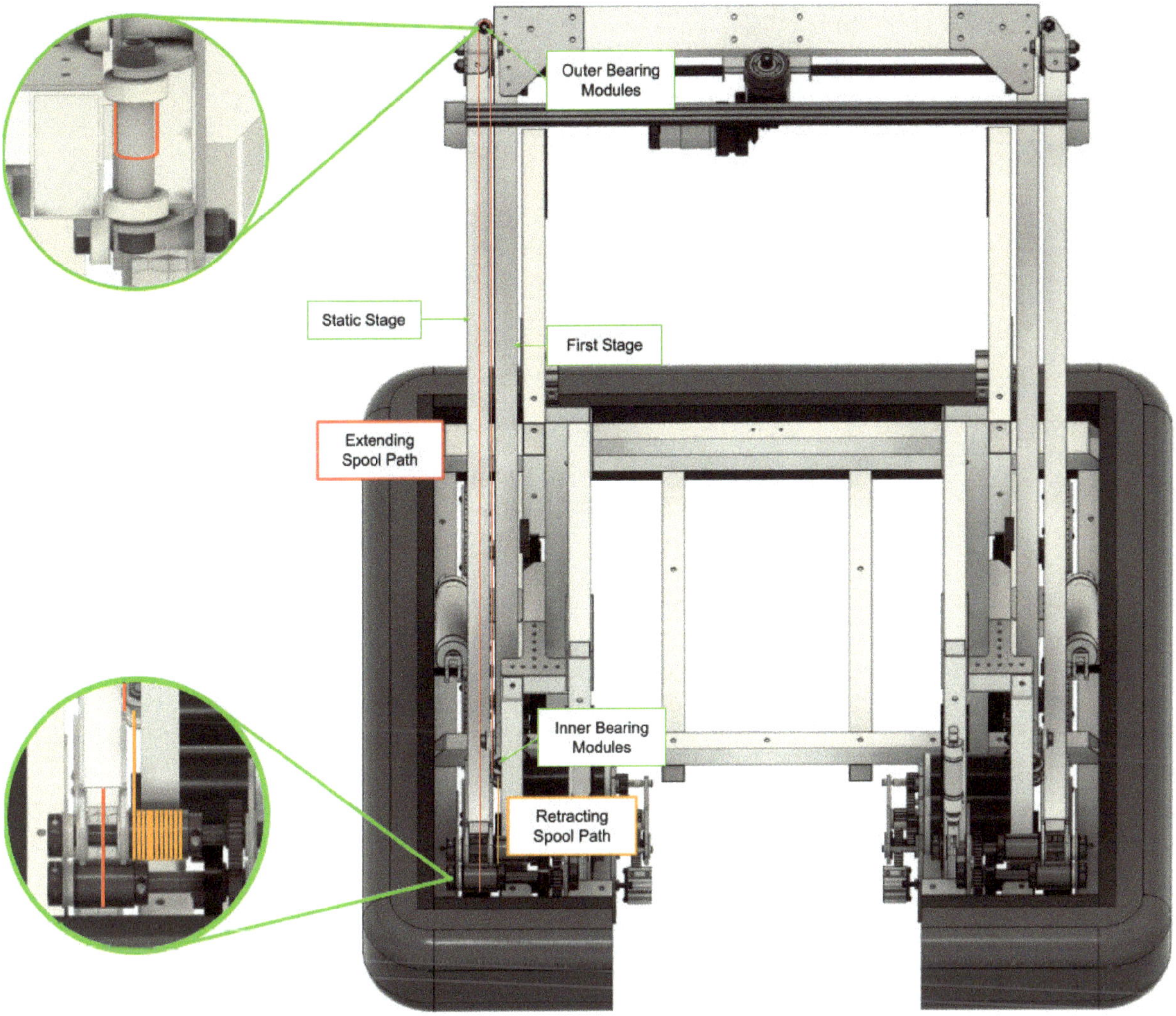


Diag. 8.6: Single Stage Lift Ex. 2 Zoomed In View; Ref. 11

Everything shown on the left side of the robot in diagram 8.6 is mirrored on the right side of the robot. A double spool lift works best when it is powered on both sides. In general, a rope rigging style for a single stage lift makes sense only if there are space constraints, since the alternate way of powering a single stage lift is simpler. It is also possible to replace the rope with a slightly flexible belt to solve some potential problems that will be explained shortly.

Note, overly flexible material should be avoided when replacing the rope. Such materials tend to weaken over time, which can make the positioning of the lift less predictable. In the worst case, such materials may break, rendering the lift unusable.

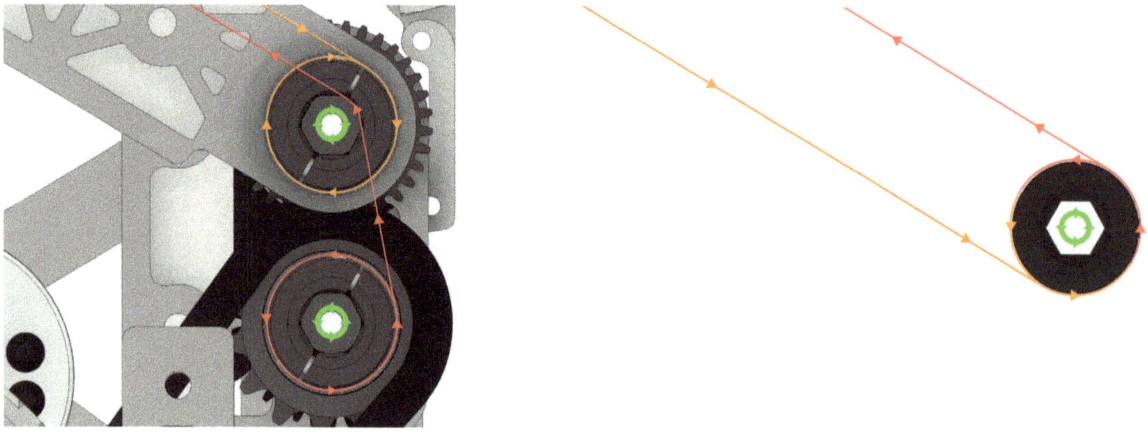

Diag. 8.7: Single Stage Lift Ex. 2 vs. Abstract Spool Path; Ref. 11

While this design uses 2 spools, (as shown in the left half of diagram 8.7), if the spool is of sufficient width to fit both ropes on, it is possible to run everything off a single winch, (as shown in the right half of diagram 8.7).

This lift's design is not ideal when it comes to limiting the travel of the elevator. With how the design works, either bearings collide with structural elements or the spools run out of rope to limit how far the lift can travel. While these limiters do function, they are not as good as having end stops between plates on the first stage and static stage as shown in the previous example of the single chain single stage lift.

Bearings ramming into structural elements is not ideal. The wear and tear on the bearing module over time will reduce contact quality between the stages and lead to rough and unreliable lift movement.

Straining the rope is also not ideal because of general wear and tear. If the rope snaps, the elevator will no longer work. If there is no spare rope available, the lift simply isn't fixable at competition.

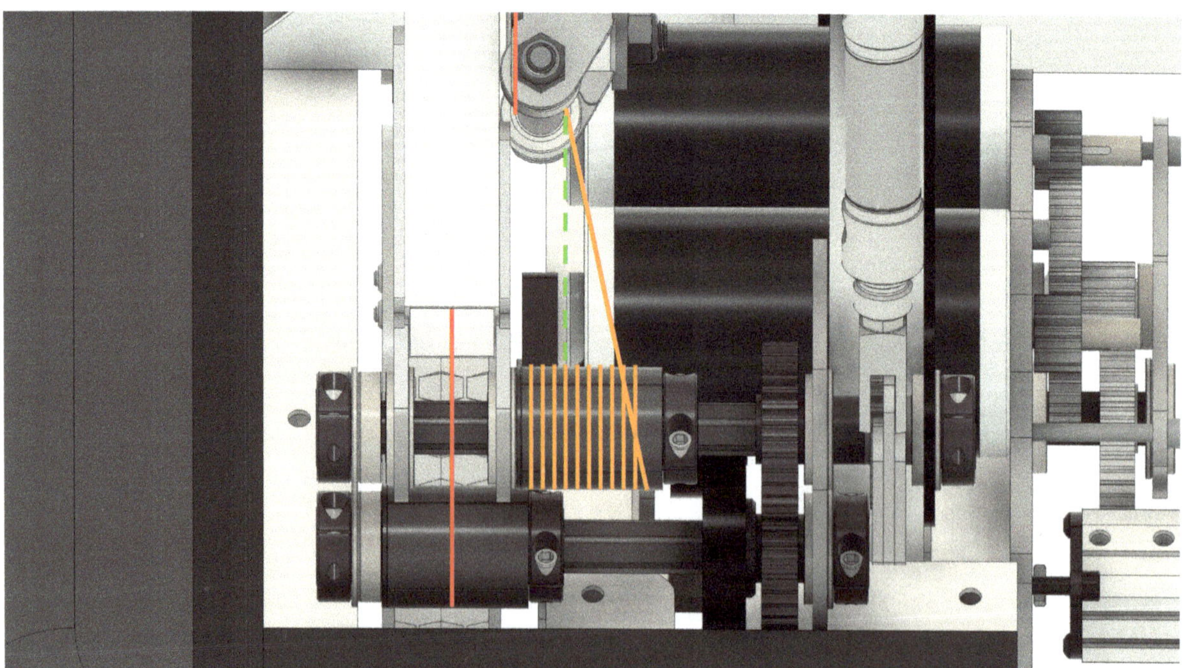

Diag. 8.8: Single Stage Lift Ex. 2 Spooling Rope Path Variation; Ref. 11

Due to the nature of one spool being wound up while the other spool is "nearly empty", there will be some expected variation in tensioning of the rope. As shown in diagram 8.8, while the red spool is fully let out, the orange spool could potentially be wound up so that the rope is not following the shortest path, shown by the dotted green line, resulting in variation of tension between the rope. (Similar variation in tensioning can occur really at any position since at least one of the spools will have some rope wrapped around it, resulting in the rope possibly not following the shortest path.) Likely, this design had some slack between the 2 spools of rope to get around this problem. This means the position of the elevator would not fully follow the movement of the spools, but this prevents the ropes from pulling on each other and minimizes the risk of the rope snapping.

Using a slightly flexible belt can mitigate some of these tensioning issues.

A more effective way to compensate for this problem is to mount the end of the rope to a spring to vary the tension on the rope. This allows for some variation in angle between the rope and the ideal centerline of the winch, while making the position of the elevator responsive to even slight changes in the position of the spools since there is no slack.

Another way to compensate for the same problem is to omit the retracting spool. The first stage will still move downwards due to gravity. However, as the lift would not be able to pull downwards, it cannot be used to climb.

To ensure proper spooling of the rope so that the rope does not overlap, the fleet angle of a rope design should be kept as low as possible. (The fleet angle is the angle between the rope's centerline as it approaches the spool and the line perpendicular to the spool's axis of rotation.) This is an issue to be dealt with in micro design.

### Double Chain Single Stage Lift

The third method of constructing a single stage lift is to use 2 chains to drive the lift. This avoids the problems that come with using rope, while also overcoming the space constraint that comes with running a chain in the center of the lift.

Sometimes due to specific space constraints it is not possible to use chain, only rope – since rope is thinner than standard #25 chain.

## 8.3 Continuous Lifts

A continuous lift is the first type of multi stage lift covered in this chapter. It takes the 2 spool design for a single stage lift and extends it to multiple stages. The basic idea is to route the extending spool of rope along a path that allows for multiple stages to be pulled up while the retracting spool connects directly to the final stage.

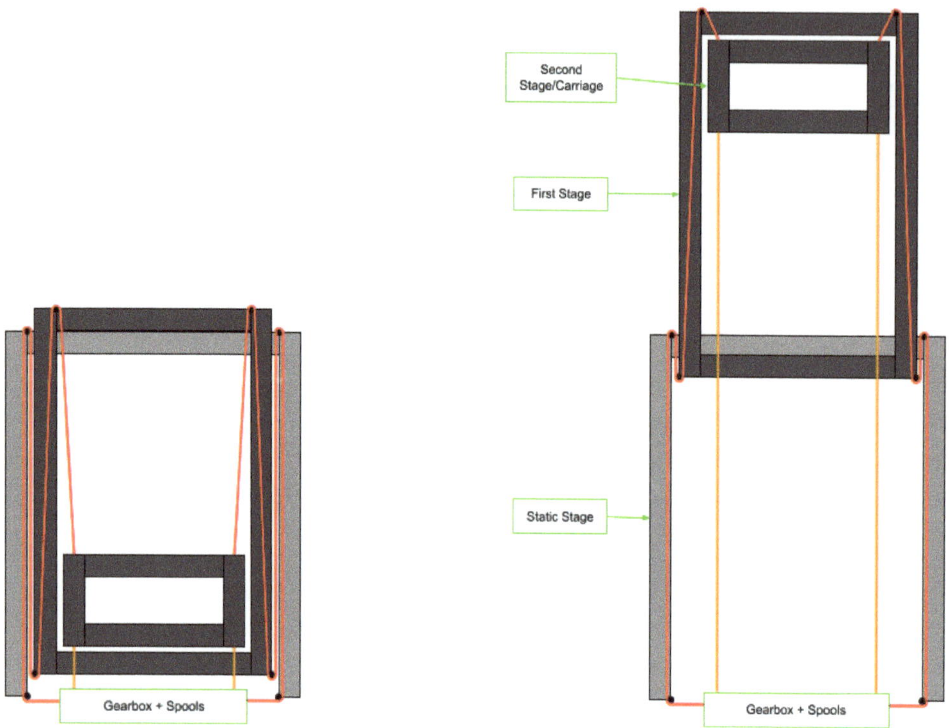


Diag. 8.9: Continuous Spool Lift Abstract

The red line represents the extending rope path, while the orange line represents the retracting rope path. The black dots represent pulleys that guide the rope to wrap around certain points on the lift. The light gray structure represents the static stage, while the darker gray structure represents the first stage and the second stage/carriage. The left side of diagram 8.9 shows the lift's down position, while the right side of diagram 8.9 shows the up position.

This idea can be further extended to support a three stage lift by adding an extra stage between the first stage and the carriage. The design could be taken even further to feature a 4th stage etc., though this should not be necessary for any normal FRC game.

All of the problems and considerations present in a single stage rope lift also apply to a continuous lift. Some problems are perhaps magnified. As can be seen in diagram 8.9, the angle slant between the top pulleys and the carriage changes between the up and down positions. This can introduce even more variation in the tension of the extending and retracting lines of rope. These problems can be compensated using the solutions outlined in Subchapter 8.2 such as omitting the down line or replacing the rope drive with a belt system.

Diag. 8.10: Continuous Lift Ex. 1 Movement; Ref. 27

The range of motion that can be achieved by a continuous lift is shown by diagram 8.10. To see an animated version, refer to https://www.frc-handbook.com/vol1/8/d10.

Given the rope path of a continuous spooled elevator, it is possible for either stage to move upward first. Usually, the second stage or carriage will move first since in an ideal system the rope path causes the net force on the first stage to be 0 until the carriage reaches the top of the first stage. However, if in reality the second stage system has a lot of friction, then the first stage may move first. (This is unlikely to be the case, since good bearing module design will ensure near 0 friction between the different stages. Another way to compensate for this is to use constant force springs on the second stage to guarantee the second stage always moves first.) It is generally better for the second stage to move first since this allows the robot to maintain a lower center of gravity. This also means the lift will move slightly faster in the bottom half of its range as the carriage on its own is lighter than the carriage with the first stage.

It is more common for an end effector like a roller claw to be attached to a multistage lift rather than a single stage lift. This is because, in order to maximize reach, at least 2 stages are required. The first stage extends beyond the starting height restriction for an FRC game, (or exists to maintain a lower center of gravity when the lift is retracted), while the second stage, in the form of a carriage, travels the distance of the first stage. This arrangement allows the claw to travel from the bottom of the lift to the very top as opposed to a more limited range of motion in the single stage lift.

Refer to https://www.frc-handbook.com/ref/28 for an example of a continuous lift from Team 2910.

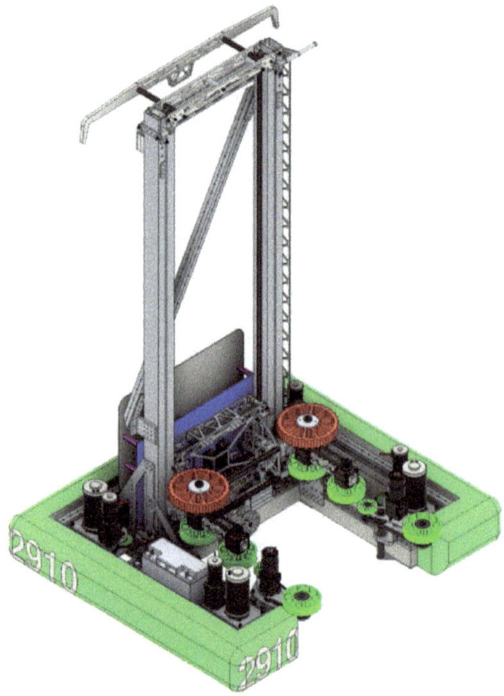


Diag. 8.11: Continuous Lift Ex. 2 Isometric View; Ref. 28

This is a continuous lift that functions as both a climbing mechanism and as a vertical lift for the claw of the robot.

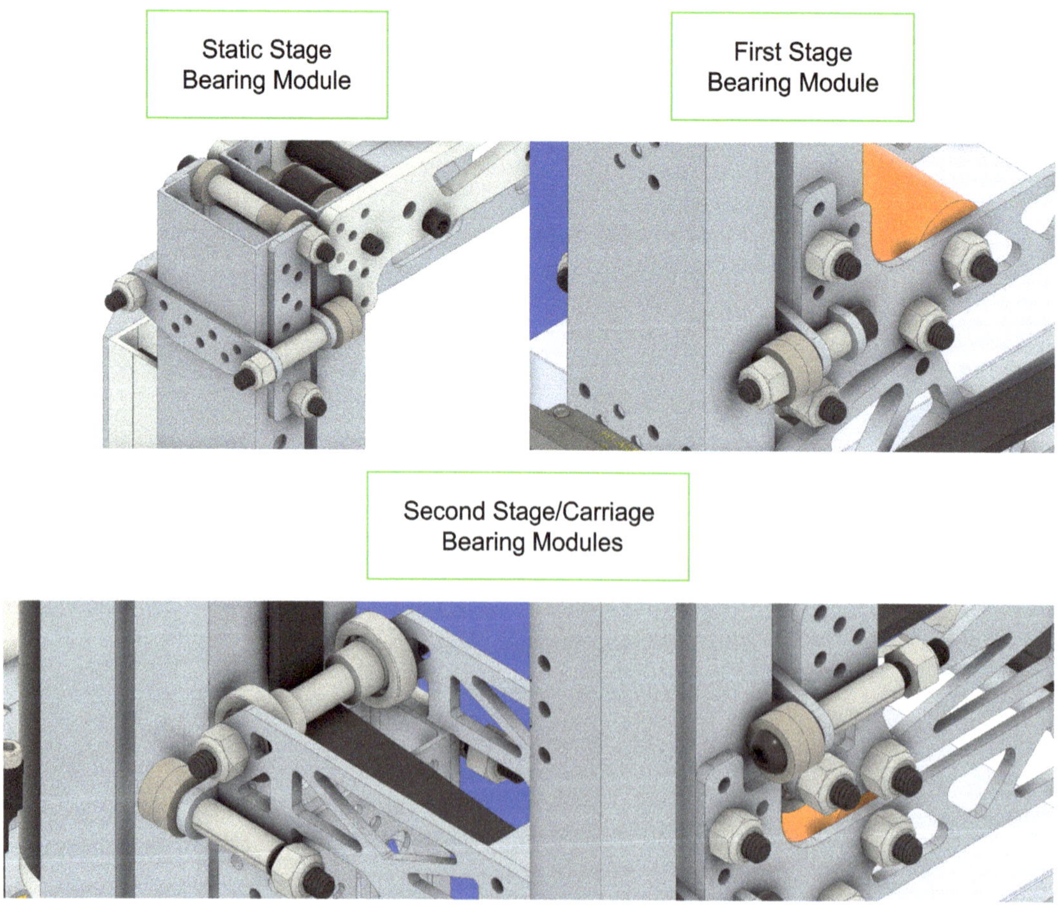


Diag. 8.12: Continuous Lift Ex. 2 Multi View; Ref. 28

This lift uses custom designed bearing modules to create a smooth motion between all of the stages. As is shown in diagram 8.12, the structure of a 2 stage lift generally follows a single stage lift, with a carriage added inside the first stage.

For lifts with even more stages, the additional stages would have bearing modules that follow the design of the static stage and first stage, with the final stage still being a carriage.

Diag. 8.13: Continuous Lift Ex. 2 Belt Path; Ref. 28

This example uses a belt as the driving system instead of a rope. The blue highlighted lines show the belt's path on each side of the robot. Instead of using 2 separate belts to control the extension and retraction of the lift, this design uses a single belt by effectively connecting the 2 belts together. This works because as shown in the other examples, when one line is reeled in, the other line is let out.

Diag. 8.14: Continuous Lift Ex. 2 Gearbox; Ref. 28

Since this design uses a belt, the belts can be directly driven by pulleys. The highlighted blue pulleys drive the belt in either direction to propel the lift up or down. The gearbox uses a combination of gears and additional pulleys and belts to drive the two key pulleys. Pneumatic friction brakes are also part of the design, used to lock the lift in position when climbing.

While this example shows it is possible to use a belt instead of a rope to make a continuous lift, it should be noted that it can be harder to get the tensioning of a belt correct than tying a string to a rotating mounting point, (like a winch). If the belt slips over the pulley as it might under high torque applications, the lift breaks. Considering this lift is capable of carrying the full weight of the robot when climbing, Team 2910 managed to effectively tension the belt. They achieved this by maximizing the contact area between the pulleys and belts and pulling on the ends of the belts during mounting, thus reducing the risk of slippage under load.

A belt is slightly more flexible than a rope. This solves some of the problems that come with adjusting angles between the various pulleys as the carriage moves. This lift is also designed in such a way that the belt's paths are as straight as possible to minimize this change in angle as the carriage move.

Diag. 8.15: Continuous Lift Ex. 2 Side View; Ref. 28

The energy chain used is highlighted in blue. Energy chain is necessary in this design because of the wiring that is required for the claw that is attached to the lift's carriage.

While the examples shown feature only 2 stages, the continuous lift concept can be extended to include 3 or more stages.

## 8.4 Cascade Lifts

A cascade lift is a different type of multi stage lift that uses a unique rigging arrangement to force all stages to move concurrently.

To start, the cascade lift uses a single stage chain lift and adds on extra stages.

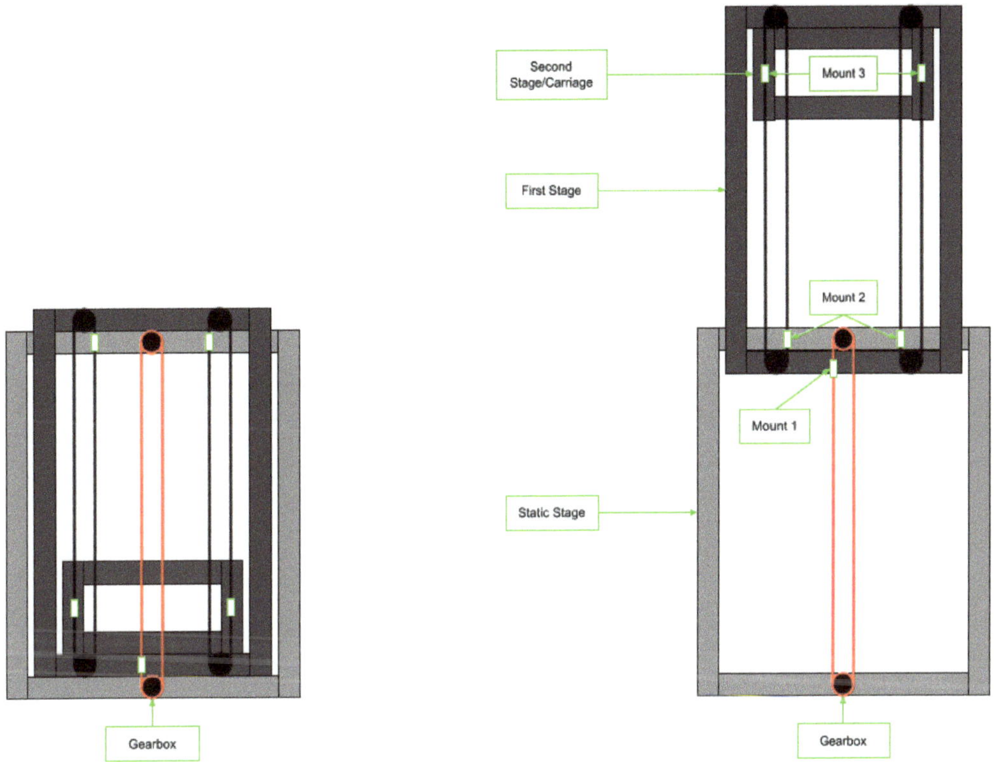


Diag. 8.16: Cascade Lift Abstract

The red lines around the 2 central axles represent the single chain driven by a gearbox at the bottom of the static stage. The chain itself is connected to the first stage shown by mount 1. The black lines around the side axles represent the pulley and belt systems that are on the first stage. The second stage/carriage is driven by the movement of the pulleys mounted on the first stage by cascade rigging. A key thing to note is that the belts are mounted at 2 locations. Mount 2 is where the belts are mounted to the static stage, while mount 3 is where the belts are mounted to the second stage. As the chain raises the first stage, the pulleys mounted on that stage start to move up too. This causes the pulleys to move relative to mount 2, which in turn drives the belts, and pulls the carriage up by mount 3.

This might be a bit hard to visualize, so here is a more detailed explanation: For the belt on the left half of the elevator in diagram 8.16, as the pulleys move up, they also move away from mount 2. This motion forces the belt to rotate clockwise around the pulleys. Since the carriage is attached to the moving belts, the carriage is now pushed upward. The same process is happening on the right side of diagram 8.16 in a mirror image.

Another way to think about it is that mount 2 and mount 3 are rigidly fixed to specific points on the same belt, and must maintain a constant distance away from each other along the belt's path. When the first stage rises, mount 2—anchored to the static stage—stays in place while the pulleys on the first stage move up. As the pulleys move up, mount 3 keeps its relative distance from mount 2 along the belt's path by moving at double the speed of the first stage. This in effect drags the carriage up at double the speed of the first stage's motion.

Due to the nature of a cascade lift, the mechanical advantage of each subsequent stage of the lift is halved. This means the torque the gearbox exerts on each subsequent stage is halved while the effective speed of the stage is doubled. A way to think of this is in terms of conservation of mechanical power. Each subsequent stage moves at twice the speed of the previous stage as the previous stage travels at a certain speed, while the next stage travels at the previous stage's speed times 2 due to the cascade rigging system. As the stage moves at double the previous stage's speed, it must halve the torque exerted on it to maintain conservation of mechanical power. This has implications for the torque calculations for a cascade lift's gearbox as well as how a lift moves.

The result is that if the exact same gearbox is used to power both a 2-stage continuous lift and a 2-stage cascade lift, the cascade lift's final stage would move at double the speed. However, because it has only half the torque, it can only lift half the weight. In reality, both lift systems would be designed to maximize speed while reaching some minimum torque for their final stage. This means a cascade lift typically requires a gearbox with a higher torque output to compensate for the reduction in lifting force.

Diag. 8.17: Cascade Lift Ex. 1 Movement; Ref. 29

The range of motion that a cascade lift can achieve is shown in diagram 8.17. Refer to https://www.frc-handbook.com/vol1/8/d17 to see an animated version of diagram 8.17. Both stages move in sync as described per abstract diagram 8.17.

Note that while most cascade lifts have 2 sets of belts and pulleys to raise the carriage, it is possible to do so with only a single set. 2 sets are typically used to distribute the load across both sets so as to stabilize the carriage.

Rope is more commonly used than smooth belts and pulleys in the cascade rigging. Technically it is possible to use sprockets and chain for the cascade rigging, but this does not really make sense since it adds additional weight compared to the belt and pulley option without a clear advantage.

A chain is usually used for the power transmission that drives the first stage upward since that is a high torque application. However, with proper tensioning, rope or belts and pulleys can also be used for the drive system.

The idea of a cascade lift can be extended to 3 or more stages. Here is an example of a 3-stage cascade lift. See https://www.frc-handbook.com/ref/30 for more details.

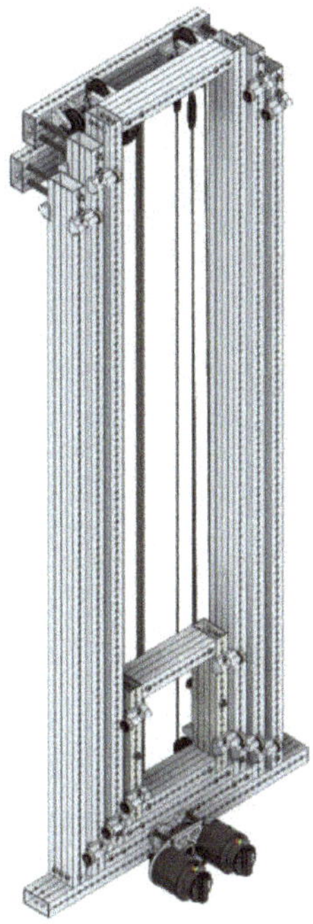

Diag. 8.18: Cascade Lift Ex. 2 Isometric View; Ref. 30

Both Reference 29 and Reference 30 are variations of a product from WestCoast Products.

## 8.5 Telescoping Lifts

A telescoping lift utilizes specialized bearing modules—that are referred to as telescoping lift bearing modules—along with constant-force springs to guide and assist in the movement of its stages.

Diag. 8.19: Telescoping Lift Bearing Module Ex. 1 Multi View; Ref. 31

These special bearing modules allow one stock to slide smoothly inside the walls of a slightly larger piece of stock. The module on the left is attached to the top of the larger stock, while the module on the right is attached to the bottom of the smaller stock.

The width of the stock used usually decreases by 0.5" between stages. As the stages get smaller, the bearing modules also get smaller. Regardless of size, the bearing modules generally follow the design shown in diagram 8.19.

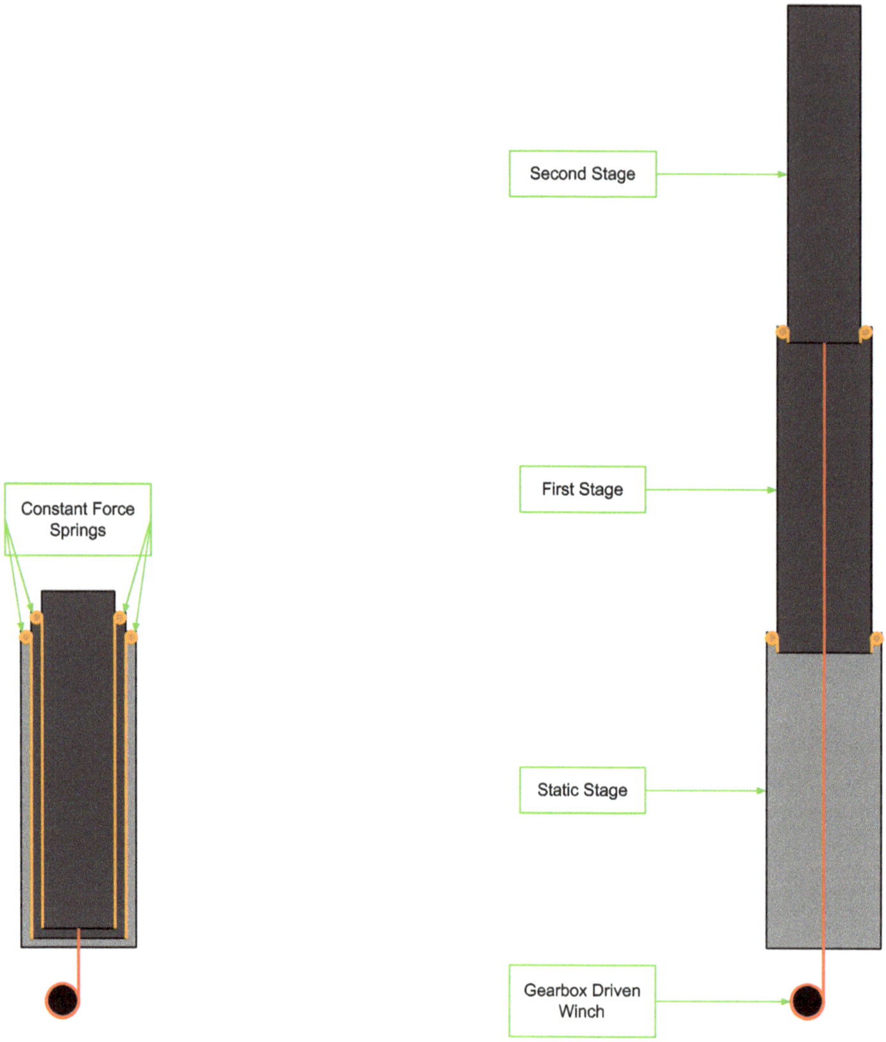


Diag. 8.20: Telescoping Lift Abstract

One variation of the single stage lift, as previously described, uses only the extending spool of string and relies on gravity to retract. Telescoping lifts can be thought of as the exact opposite. Diagram 8.20 shows the cross-section of the different stock pieces, each recessed inside the next. Constant force springs, shown as orange lines, provide a passive upward force to extend the telescoping lift. A retracting spool of string, shown by the red line, holds the lift at specific positions. Constant force springs are required between each stage to trigger the upward motion of the lift.

Diag. 8.21: Telescoping Lift Ex. Isometric View; Ref. 8

This example is of a 2-stage telescoping lift. It is used as a climbing mechanism. See https://www.frc-handbook.com/ref/8 for more information on the game and full robot.

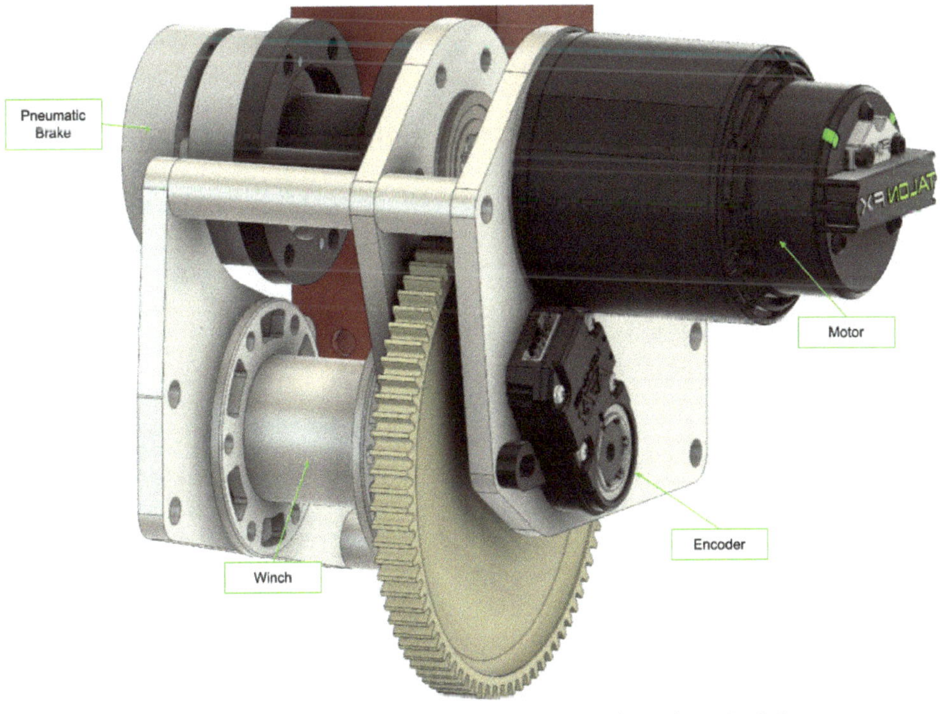


Diag. 8.22: Telescoping Lift Ex. Zoomed In View; Ref. 8

Diagram 8.22 shows a typical gearbox for a telescoping lift. An encoder keeps track of the lift's position by measuring the rotation angle of the winch. When the telescoping lift has climbed to the desired position, a pneumatic friction brake locks the winch. This transfers the strain of holding the robot in place from the motor to the brake.

Diag. 8.23: Telescoping Lift Ex. Movement; Ref. 8

The range of motion that can be achieved in this telescoping lift is shown by diagram 8.23. See https://www.frc-handbook.com/vol1/8/d23 for an animated version.

The use of constant force springs usually result in the lower stages extending first. This is because, in standard telescoping lift kits, the constant force springs used to pull up the lower stages are stronger than the ones on the higher stages. If there is sufficient friction in the lower stages, the upper stages may extend first instead.

Constant force springs are relatively weak mechanisms. As the upward motion of the telescoping lift is propelled by constant force springs, telescoping lifts— combined with light-weight static claws—are usually used only for climb mechanisms. Not much force is required to simply extend the arm. In contrast, a strong gearbox that powers the retracting winch is able to lift the robot off the ground. It is possible to add a heavier end effector, such as a gripper claw, to the end of a telescoping lift. However, it must still be relatively light because constant-force springs are only so strong.

There is very little variation in how a telescoping lift is designed, so further examples will not be shown. The main elements that change between applications are the gearbox design, the claw or end effector design, and the length of each stage.

### Telescoping Lifts vs. Lifts That Use Telescoping Bearing Modules

As a general note, a lift can use telescoping lift bearing modules without applying the telescoping extension and retraction system. Such lifts are not classified as telescoping lifts; instead, they are identified by the type of rigging system used to produce movement. An example of a cascade lift that uses telescoping lift bearing modules can be seen in the second example of Subchapter 9.3.

## 8.6 Further Comparison of Lifts/Elevators

In general, telescoping lifts are primarily used for climbing, while single-stage lifts are used for applications requiring limited reach. Continuous lifts and cascade lifts tend to be used for similar applications: supporting a heavy end effector that needs to travel vertically for a longer distance than a single-stage lift can reach. This overlap often leads teams to ask the question: Which is better—a continuous lift or a cascade lift?

A continuous lift is often a bit simpler to design, as the rope path starts at the gearbox and is intuitively routed around all of the stages before ending at the final stage. Because of this routing, springs are required to maintain tension between the 2 lines of ropes. Alternatively, the down line can be omitted, or a long, but more expensive, timing belt can be used.

A cascade lift is a bit more complicated to design due to the multiple sets of ropes involved. Furthermore, the additional torque required to move all stages simultaneously may necessitate a larger or more intricate gearbox.

In terms of execution, cascade lifts are considered the more challenging of the two. A cascade lift requires multiple cascade rigs to be connected in an aligned manner. As cascade lifts tend to be more compact, which can help meet certain design requirements, the construction difficulty level increases.

In terms of performance, a 2-stage continuous lift is typically faster in the bottom half of its movement range because it is only lifting the carriage stage. In contrast, a cascade lift moves all stages simultaneously throughout its movement range. A 2-stage continuous lift is usually predictable in both position and speed, but not entirely. Friction and rapid motion can cause the first stage to begin moving before the second stage reaches the top of the first. A cascade lift, in contrast, will be more consistent in speed and position since all stages always move together. Arguably, continuous lifts are easier to design and build, but this really depends on how much experience an FRC team has with either lift system.

## 8.7 Summary on Lifts/Elevators

Lifts can be used to manipulate many different types of end effectors, such as all of the claw types (see Chapter 10). In particular, FRC Games almost always involve a final climb as an end goal, which is usually best accomplished with a lift and a static claw (see Subchapter 10.2). Such a lift is a staple feature in almost every competition robot.

See a summary of all of the lift types used in FRC games on the next page:

<table>
<tr><th colspan="2">Single Stage Lift</th></tr>
<tr><td colspan="2"><ul><li>Simple design</li><li>Pulls a single stage up and down attached to either chain or rope</li><li>Lift is fully powered in both directions</li></ul></td></tr>
<tr><td>Single Chain Single Stage Lift</td><td><ul><li>Driving System: Chain</li><li>Chain runs down the middle of the lift for even load bearing</li></ul></td></tr>
<tr><td>Double Chain Single Stage Lift</td><td><ul><li>Driving System: Chain</li><li>2 sets of chains run down the lift, one on either side</li></ul></td></tr>
<tr><td>Double Spool Single Stage Lift</td><td><ul><li>Driving System: Rope</li><li>Rope is routed through the lift, with one line pulling the lift up and the other pulling the lift down</li><li>Down rope can be omitted, relying on gravity for retraction</li></ul></td></tr>
<tr><th colspan="2">Continuous Lift</th></tr>
<tr><td colspan="2"><ul><li>Simple design</li><li>A multi stage lift</li><li>A rope runs the full length of the lift to pull the lift up with a separate rope to pull the lift down</li><li>Upper stages are pulled up before lower stages</li><li>Lift is usually fully powered in both directions</li><li>Down rope can be omitted, relying on gravity for retraction</li><li>Driving System: Rope/Belt</li><li>Potential problem with rope tensioning due to long rope length. Solutions:<ul><li>Use springs at attachment point (watch for added weight)</li><li>Omit down rope (may reduce retraction reliability)</li><li>Use a belt (long timing belts = higher cost)</li><li>Retie ropes frequently</li><li>Switching the rope to a belt which is more flexible.</li></ul></li><li>Mechanical Advantage: Consistent stage to stage</li></ul></td></tr>
</table>

| Cascade Lift |
| --- |
| • Complex design<br>• A multi stage lift<br>• Multiple ropes, belts and/or chains that work together to move all stages simultaneously<br>• All stages move in sync<br>  ○ Requires higher gear ratio than continuous lift due to moving multiple stages at once<br>• Lift is fully powered in both directions<br>• Driving System: Chain/Rope<br>  ○ Rope tensioning is more easily managed than in a continuous lift as it only routes through one stage<br>• Cascade Rigging System: Rope/Belts<br>  ○ Cascade rigging is usually done on either side of a lift to better stabilize the lift and distribute the load<br>• Mechanical Advantage: Speed is doubled and torque is halved for each additional stage<br>  ○ Requires a very high torque gearbox due to this |
| **Telescoping Lift** |
| • Complex design (Simple if bought as a set)<br>  ○ A single rope holds the lift down while constant force springs passively pull the lift upward<br>  ○ Very compact: stock pieces of 0.5" in difference are recessed inside each other<br>• Lower stages tend to extend before upper stages due to the difference in spring strength between stages<br>• Lift is fully powered in the down direction and minimally powered in the up direction<br>  ○ Lift is effective at pulling a robot off the ground, but is not effective at pulling an end effector, such as a claw, up<br>• Mechanical Advantage: Consistent stage to stage |

Diag. 8.24: Lifts/Elevators Summary Table

The following flow chart highlights the factors typically considered in the selection of lift type for a robot design. However, since design priorities differ from team to team, the order of the factors to be considered may also differ. Therefore, this flowchart may need to be adjusted to suit each team's requirements. Note, this flowchart does not encompass all methods of end-effector manipulation, which can also be achieved using arms (see Chapter 7) or multi-axis arms (see Chapter 9).

"Reach" in the flowchart refers to the maximum height of the lift when fully extended. Lifts are typically designed to extend from the maximum allowed starting dimension of the robot. More stages result in lifts with higher reach.

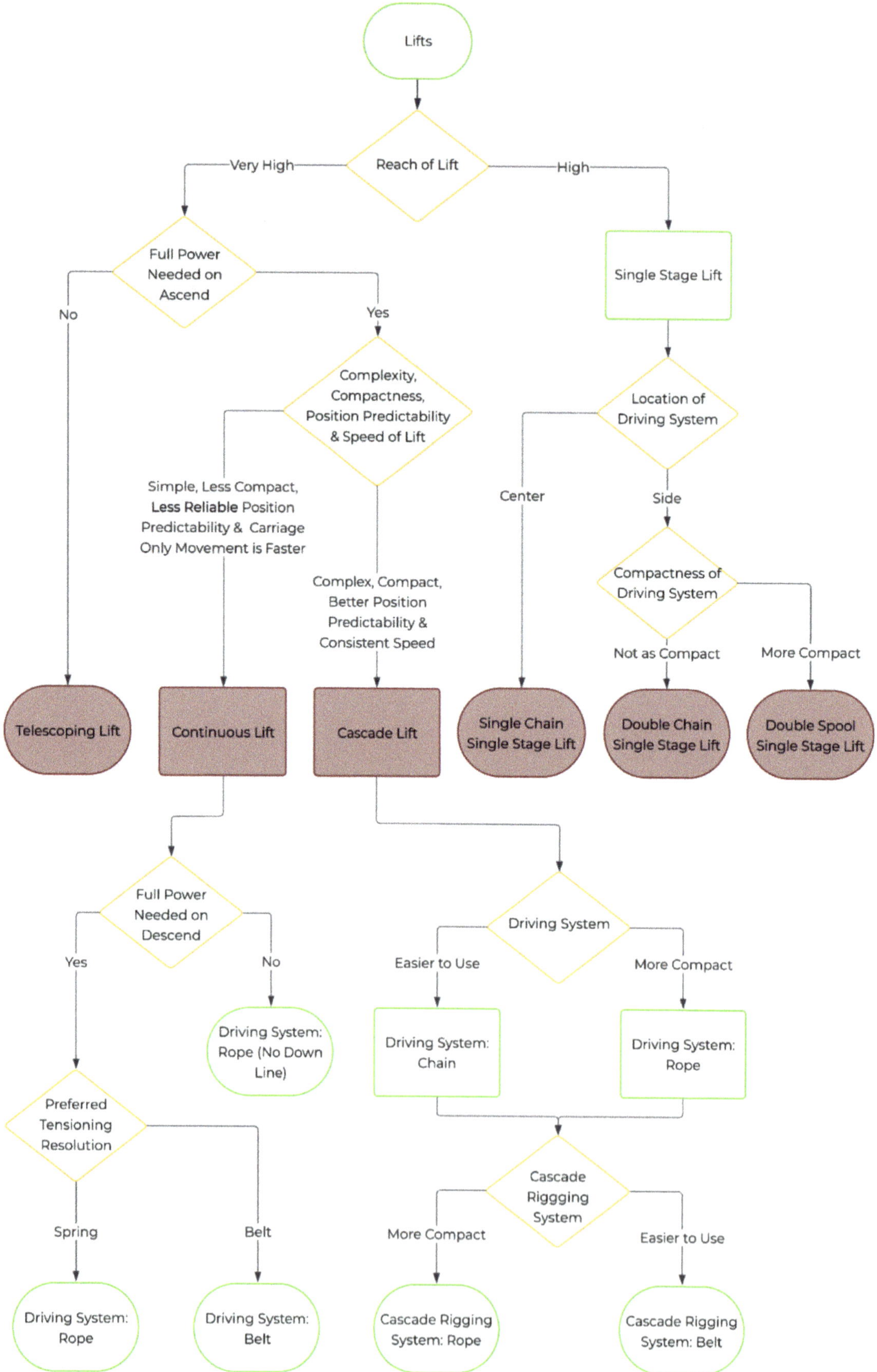


Diag. 8.25: Lifts/Elevators Flowchart

# 9. Advanced: Multi-Axis Arms

## 9.1 Introduction

The different advanced arm and lift mechanism combinations described in this chapter are built upon the same idea: to incorporate multiple axes of movement (both rotational and linear) into a single mechanism. For this reason, all of the mechanisms described in this chapter can be classified as multi-axis arms.

Here are some ways advanced FRC teams increase the flexibility and/or reach of arms and lifts:

1. Arms on Lifts: An arm is attached to the final stage of a lift. (See Chapter 8 on Lifts/Elevators.)
2. Extendable Arms: Arms with embedded lift mechanisms that allow for the arm to extend and retract.
3. Multi-Jointed Arms:. Pivot arms or chain bar arms are stacked on top of another, treating the previous pivot arm or chain bar arm as the current arm's "tower."
4. Turntable Addition: Any arm, lift, or multi-axis arm can be placed on a turntable, allowing said mechanism to face any direction relative to the drivebase. (See Subchapter 5.7 on turntable considerations.)

Generally, the more advanced multi-axis arms require complex calculations to determine the optimal gear ratio for their intended range of motion and load.

Multi axis arms are similar to arms as they are often used to move some other mechanism in a robot, typically a claw. As has been previously mentioned, the mechanism being moved is referred to as the end effector.

The following subchapters cover examples of these more advanced arm types.

## 9.2 Arms on Lifts

Mounting a pivot arm on a lift increases the arm's flexibility.

The example on the next page in diagram 9.1 shows a chain bar arm on a single stage lift. (For more information on the game and full robot see https://www.frc-handbook.com/ref/32.)

Diag. 9.1: Arm on Lift Ex. Isometric View; Ref. 32

Here's a brief description of how the lift works.

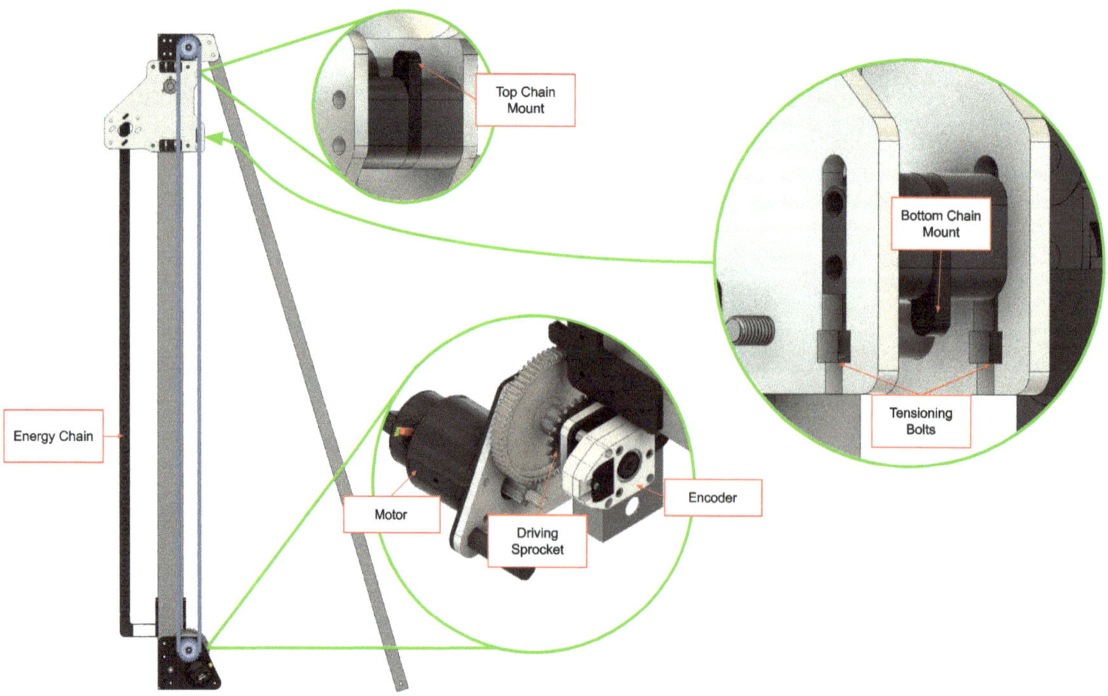


Diag. 9.2: Arm on Lift Ex. Zoomed In Views; Ref. 32

This lift uses a double chain system to drive the carriage or first stage. The chain follows the light blue outline shown on the far left of diagram 9.2. The gearbox zoom in shows the gear reduction that is used to drive the sprocket that turns the chain. The position of the chain is measured by an encoder mounted on the same axle. The top chain mount, shown in the top zoom-in view, combines laser-cut Delrin and 3D printed parts to create a #25-chain compatible mount. The bottom chain mount works the same way, but is mounted on a slide with slotted holes for tensioning bolts. The chain can be properly tensioned by tightening the tensioning bolts. Everything shown is mirrored on the other side, making this a double chain single stage lift.

The far left also shows an energy chain that houses all of the wiring for the arm and the intake. An energy chain provides effective wire management for a moving carriage.

Diag. 9.3: Arm on Lift Ex. Zoomed In Views; Ref. 32

This chain bar uses 4 sprockets, highlighted in blue, to create the parallel motion characteristic of a chain bar linkage. The sprockets in the top right are mounted statically to the inside of the carriage, while the sprockets on the bottom left are mounted statically to the outside of the intake.

The top of the carriage is left open so that the chain-bar arm is able to wrap all the way around the robot if required during game play. Normally it would be a bad idea to rotate such a long arm around the entire robot since the center of gravity of the robot would be raised significantly, making it unstable. However, the pivot of the arm can be lowered to the lift's minimum when performing this action, making the raise in center of gravity a lot more manageable.

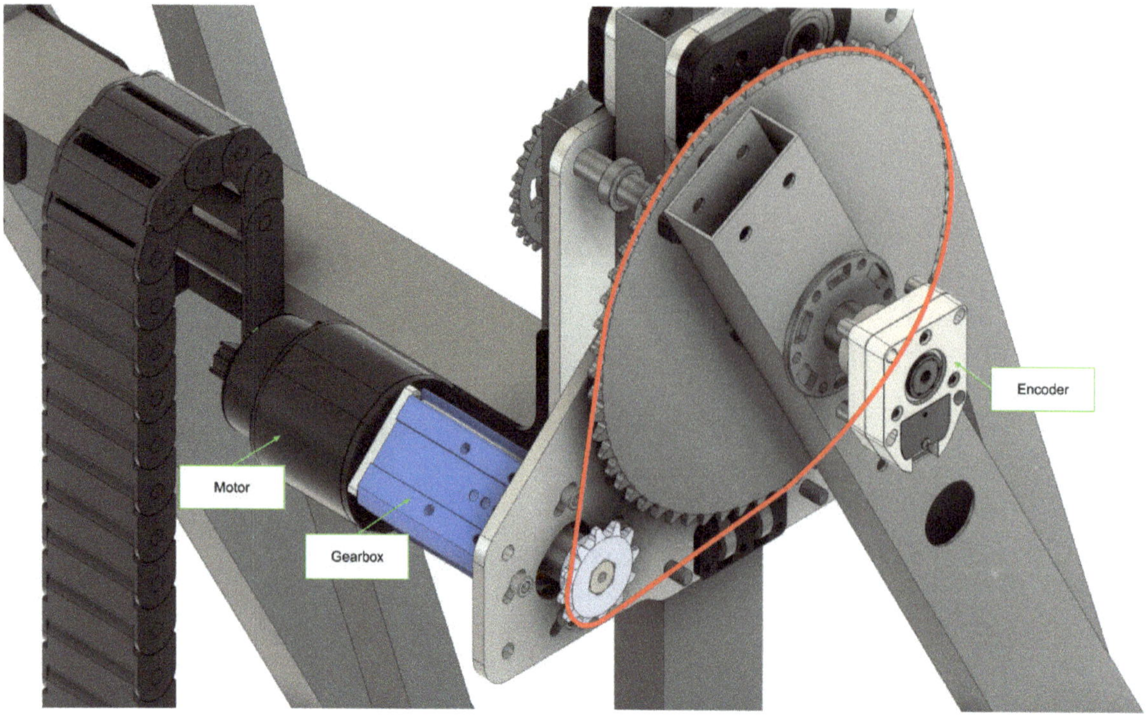


Diag. 9.4: Arm on Lift Ex. Zoomed In Views; Ref. 32

To power the arm, the motor first uses a sport gearbox to reduce the gear ratio significantly. The gear ratio is further reduced through the use of a small sprocket from the output of the gearbox to drive the larger sprocket, shown by the red line, which is meant to represent a #35 chain. The large sprocket is mounted directly to the arm which allows the sprocket to stably turn the arm. An encoder on the final output axle measures the rotation of the chain-bar arm. The plate the encoder is mounted on is hidden from view so that the interaction between the 2 sprockets can be more clearly seen.

As this is a chain system, a tensioner is required. The gearbox that drives the smaller sprocket is on slot holes so that a cam, (hidden behind the metal plate in diagram 9.4), is able to push the gearbox away from the larger sprocket to properly tension the chain.

This arm design is mirrored on the other side so that both chain bar arms can move in sync.

With the combination of a chain bar arm on a lift, the intake at the end of the chain bar can be moved to a great variety of locations. The intake can also move linearly up and down with the arm fixed at any angle. This in theory allows for a smoother pick up and release of game objects.

## 9.3 Extendable Arms

We will look at 2 examples of extendable arms. The first is a single stage extendable arm which utilizes the concept of the single stage lift and the second is a cascade extendable arm which utilises the concept of the cascade lift.

### Single Stage Extendable Arm

Diag. 9.5: Extendable Arm Ex. 1 Isometric View 1; Ref. 8

Diagram 9.5 is that of an extendable arm. It is a pivot arm with a single stage elevator system built on the arm. The range of motion of this arm has been described in earlier chapters (see Subchapters 3.2 and 5.3).

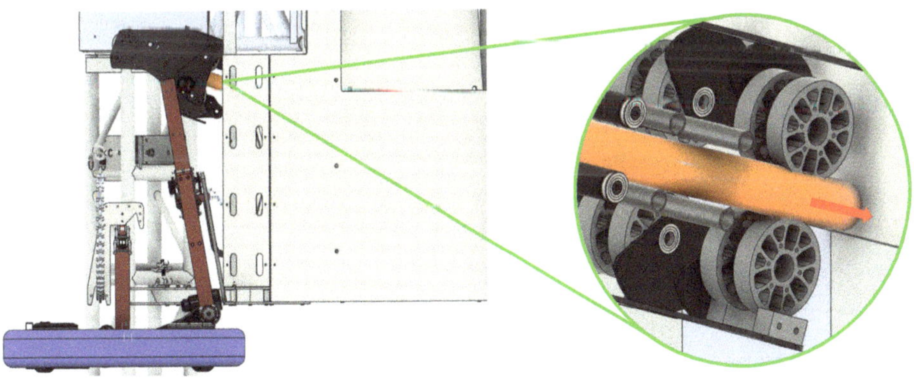

Diag. 9.6: Extendable Arm Ex. 1 Side View; Ref. 8

A positioning capability of this extendable arm not previously highlighted, (and the primary reason why this flywheel design is on an extendable arm to begin with), is that when this arm is fully extended and the robot's climb is drawn all the way down, the flywheel is able to reach the trap to score additional points during the endgame.

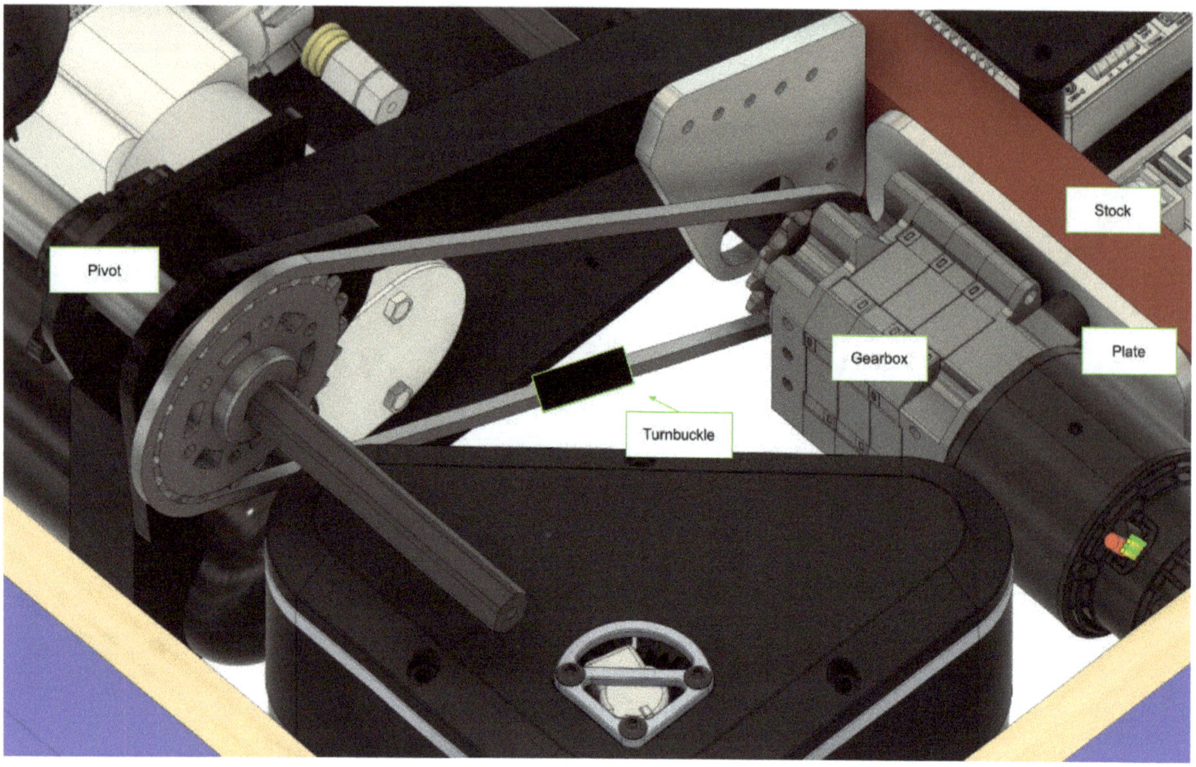


Diag. 9.7: Extendable Arm Ex. 1 Isometric View 2; Ref. 8

To explain how this extendable arm works, the pivot arm is described first. The pivot arm is driven by a chain, with the motors and gearbox positioned lower in the drivebase. This layout was chosen to work around the awkward placement of the swerve modules, which blocked the space where a gearbox would normally be built—beside the actual pivot point. The chain is tensioned with a turnbuckle, depicted by the black rectangle in diagram 9.7. Because the pivot arm is physically limited to a 180-degree range (with a programming limit that further restricts it to 90 degrees), the turnbuckle never comes close to the sprockets, making it an acceptable tensioning technique.

During initial construction, the chain's tension tore the gearbox from its mounting stock. To address this, thicker 3 mm stock was used, and a plate was added beside the stock piece to increase the gearbox's contact surface area. This reinforcement prevents shearing and keeps the gearbox securely mounted.

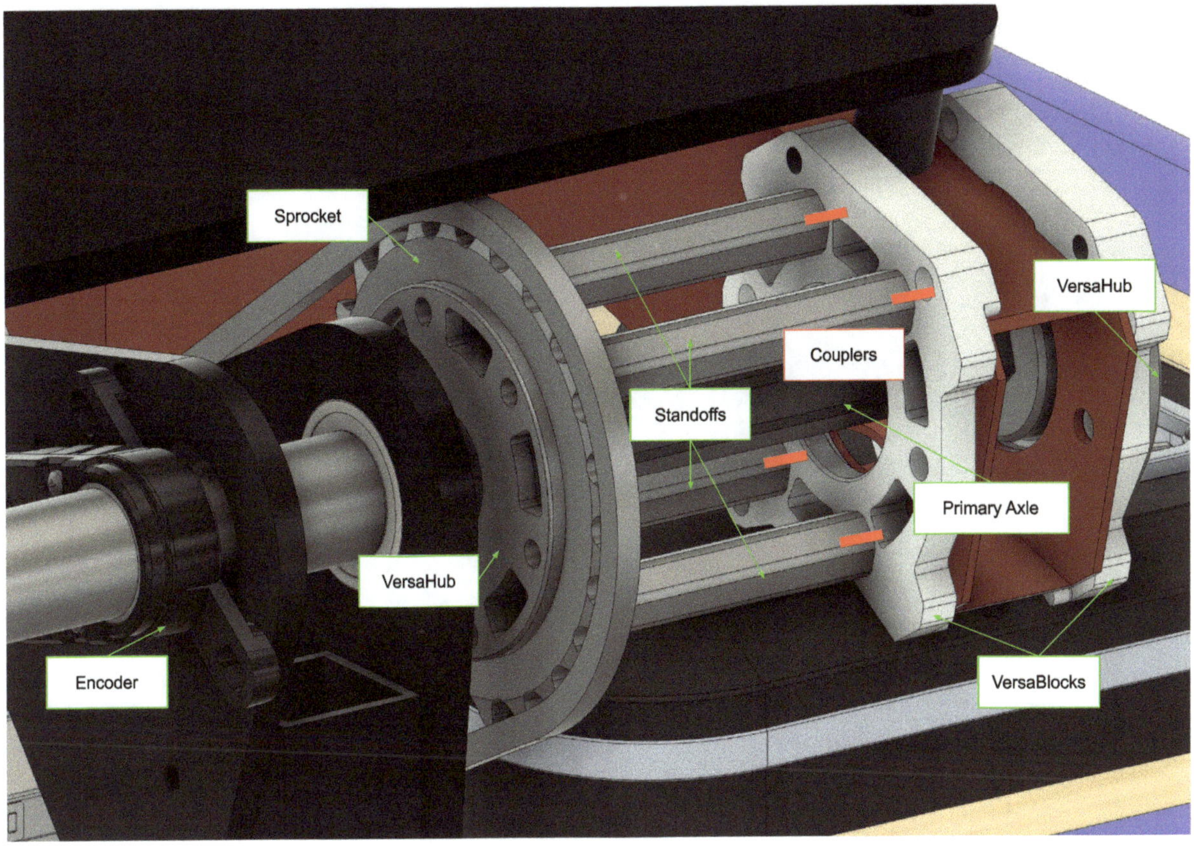


Diag. 9.8: Extendable Arm Ex. 1 Zoomed In View 1; Ref. 8

The sprocket turning the pivot arm translates motion by standoffs between the sprocket and versablock. Standoffs are connected to the sprocket by short bolts, while the other end utilizes couplers as highlighted by the red lines. The sprocket and outer versablock also has a versahub installed so that rotational motion can be translated through the primary rotational axle. This is important as the encoder can then track the arm's position by following the rotation of the primary axle.

Ideally, the standoffs should be shorter. The robot made it through the regional event with fixes and adjustments between each round, but over time the standoffs loosened and/or warped, making it harder to control the position of the arm. With shorter standoffs this problem would have been minimized, however, a wide arm system was determined to be necessary so that the intake could fit within the width of the arm.

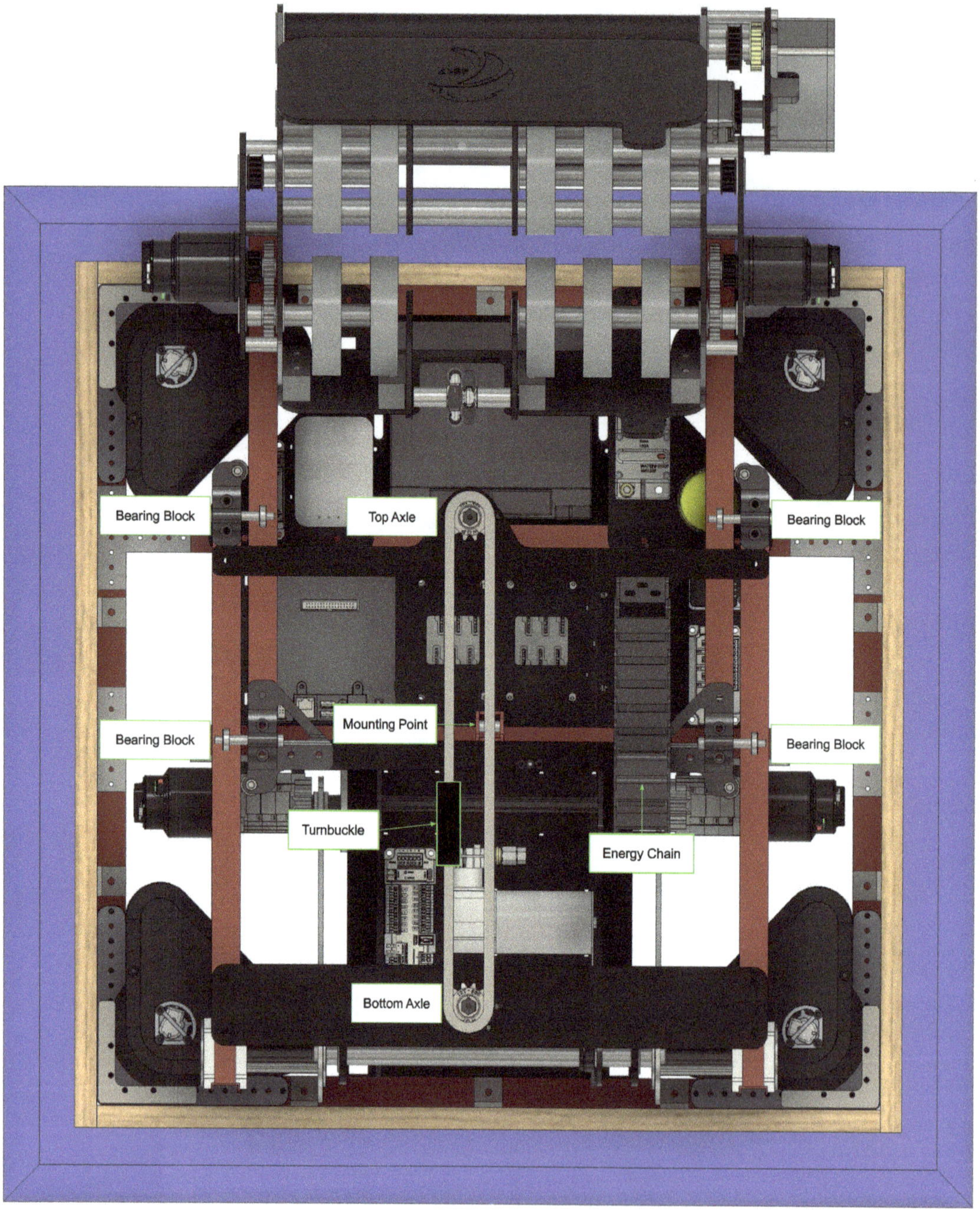


Diag. 9.9: Extendable Arm Ex. 1 Top View; Ref. 8

For the single stage lift on the arm, the design is relatively straight forward. Bearing modules are used to give the lift a smooth sliding action. A chain is connected around the top and bottom axle, with a mounting point connected to the chain which allows for the lift to be driven. A turnbuckle is used to tension the chain as this chain will never travel even half of the full distance around both sprockets. Energy chain is used to guide all of the wiring needed for the intake and flywheel assemblies at the end of the lift.

### Cascade Extendable Arm

The meta in FRC has slowly gravitated toward thin extendable pivot arms. This design is very versatile and can be used for most FRC stacking games. Refer to https://www.frc-handbook.com/ref/32 for such an example from Team 2910.

Diag. 9.10: Extendable Arm Ex. 2 Isometric View; Ref. 33

This extendable arm uses a pivot arm to turn what looks like a telescoping lift, but is actually a cascade elevator. (See Chapter 8 on Lifts/Elevators.)

Diag. 9.11: Extendable Arm Ex. 2 Side View 1; Ref. 33

The pivot arm is driven by 2 gearboxes on either side of the arm. Each gearbox houses 2 motors, making up a total of 4 motors for this mechanism. Diagram 9.11 shows the side view of one of these gearboxes. A series of gears powered by the motors turn the driving sprocket, as indicated by the bottom right red circular arrows. This drives a big chain, shown by the dark yellow arrow set, which in turn rotates the pivot, indicated by the top left red circular arrows.

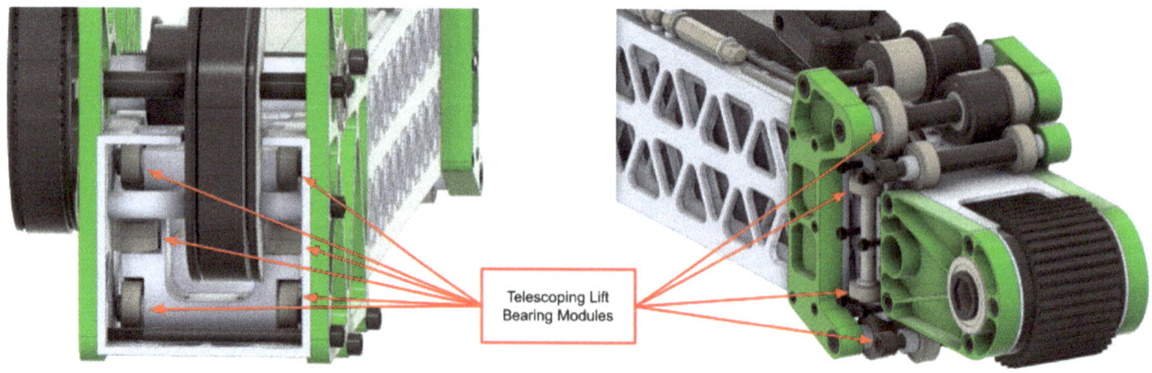


Diag. 9.12: Extendable Arm Ex. 2 Zoomed In View 1; Ref. 33

The cascade lift is a bit more complicated. Stock pieces fit inside one another and are able to move smoothly due to the telescoping lift bearing modules.

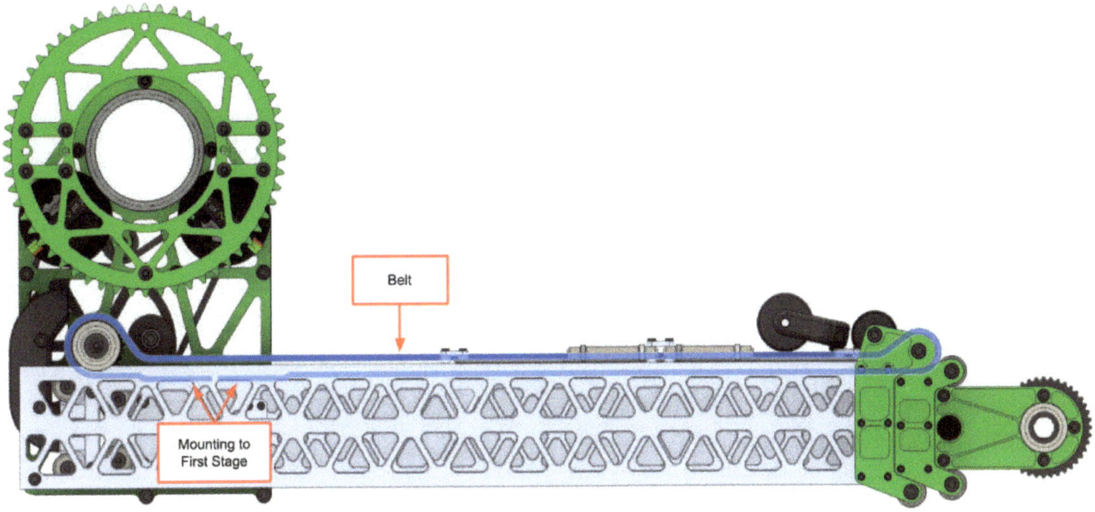


Diag. 9.13: Extendable Arm Ex. 2 Side View 2; Ref. 33

This belt performs the role of the chain in a cascade lift. Diagram 9.13 shows where the belt is mounted to the first stage of the cascade lift.

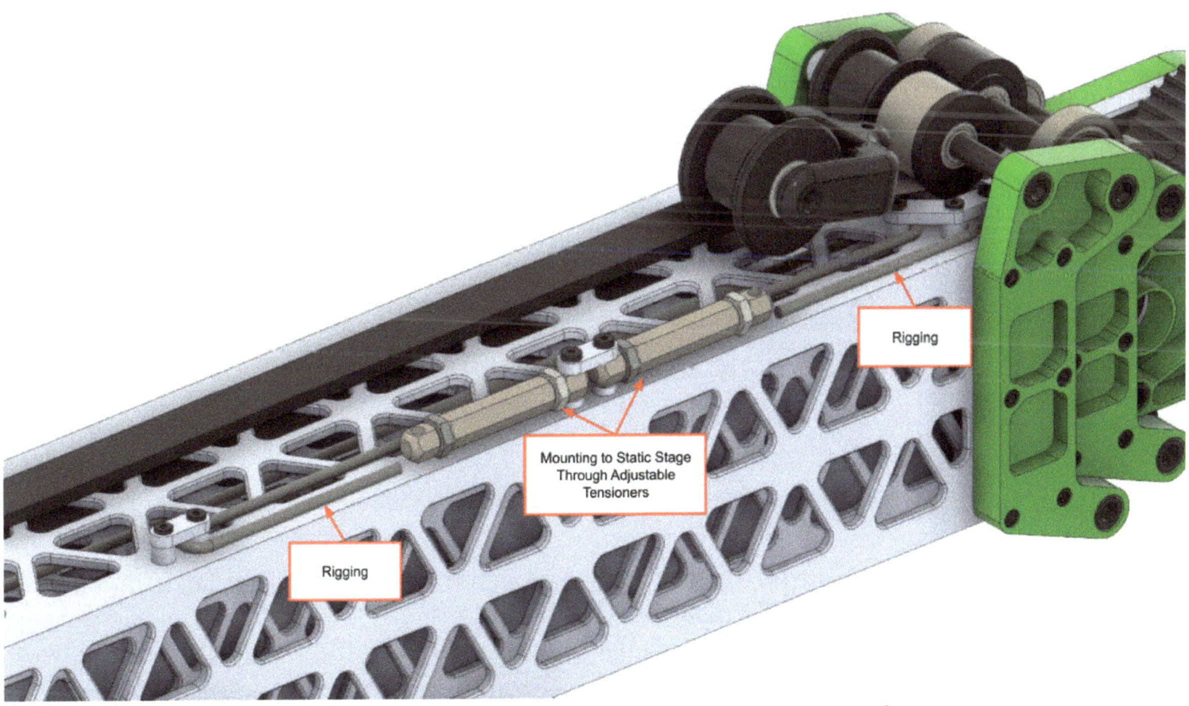


Diag. 9.14: Extendable Arm Ex. 2 Zoomed In View 2; Ref. 33

The rigging of the rope that causes the cascading action is shown in diagrams 9.14-9.16.

Diagram 9.14 shows where the rope is bolted onto the static stage.

Diag. 9.15: Extendable Arm Ex. 2 Zoomed In View 3; Ref. 33

Diagram 9.15 shows the 2 pulleys of the rigging that are embedded in the first stage of the cascade lift. While the rigging follows a complicated path due to the tight space tolerances between stocks in a telescope-like lift structure, the functional effect of the rigging is the same.

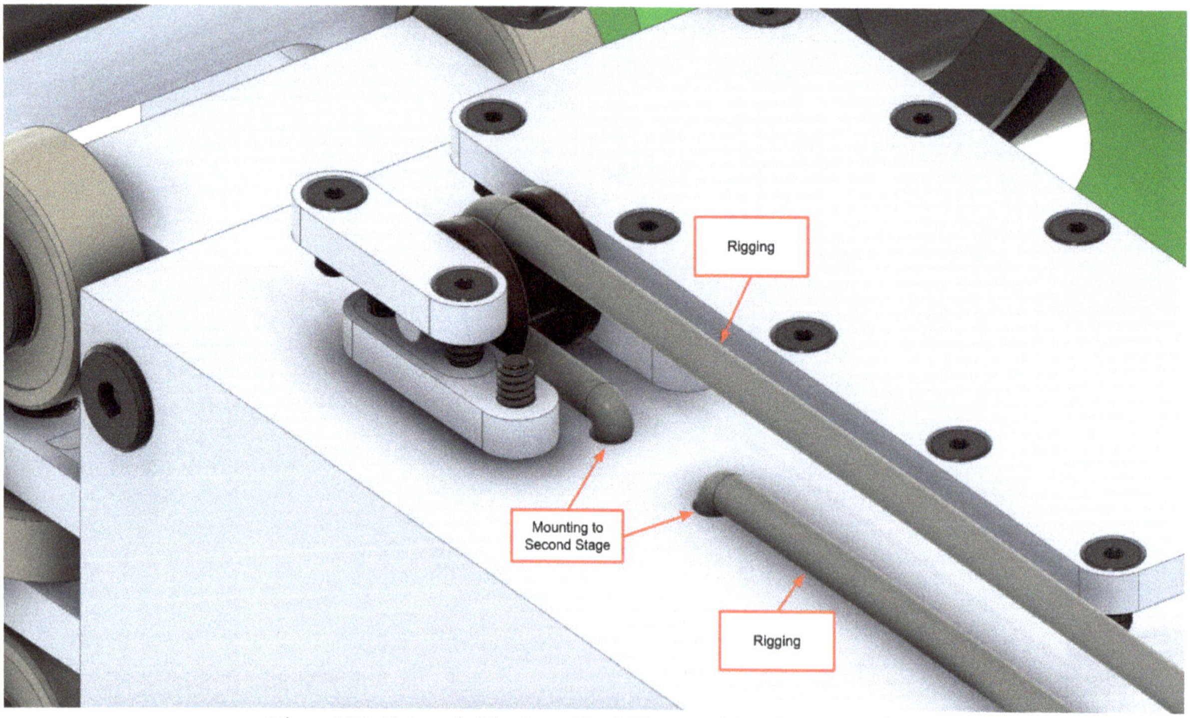


Diag. 9.16: Extendable Arm Ex. 2 Zoomed In View 4; Ref. 33

The final step is the mounting of the rope to the second stage, as indicated in diagram 9.16. This rigging, combined with the pulley's motion that moves the first stage, causes the first and second stage of the lift to move in sync, making this mechanism a cascade lift.

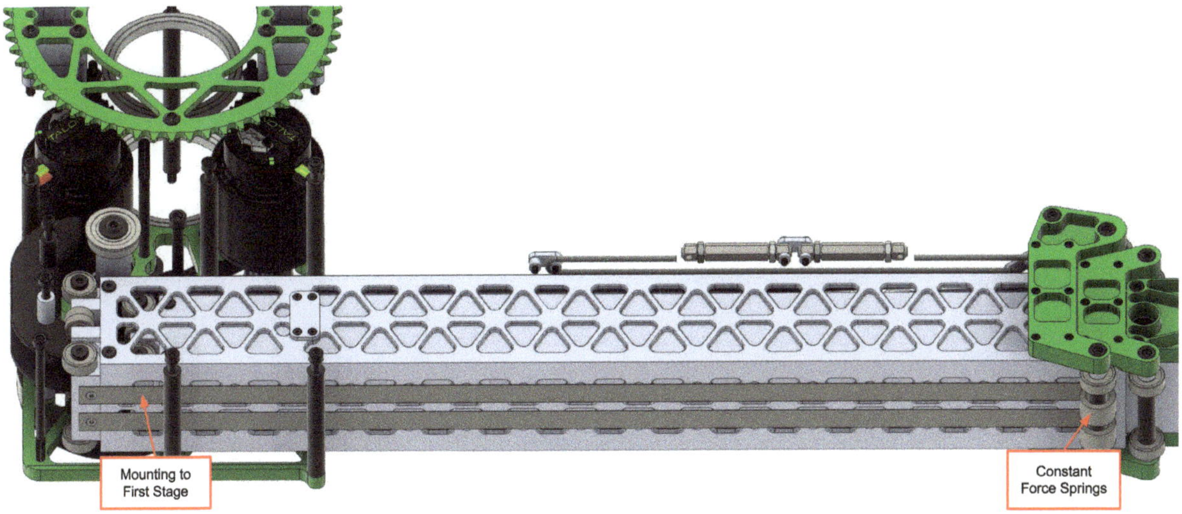


Diag. 9.17: Extendable Arm Ex. 2 Bottom Slanted View; Ref. 33

The CAD file shows an initial plan to use constant force springs to help move the first stage of the lift. Based on how this robot was used in gameplay, the arm would generally be pointing upward when extending the cascade lift. These springs would decrease the strain on the motors during extension. However, Team 2910 ultimately omitted them during the build as it was found that the lift was sufficiently strong without these constant force springs.

Another note-worthy part of this design is the wire management system, especially given that this is a very compact 2-stage lift. Team 2910 uses a clever combination of a constant force spring and custom energy chain routers to manage cables effectively on a different face of the arm.

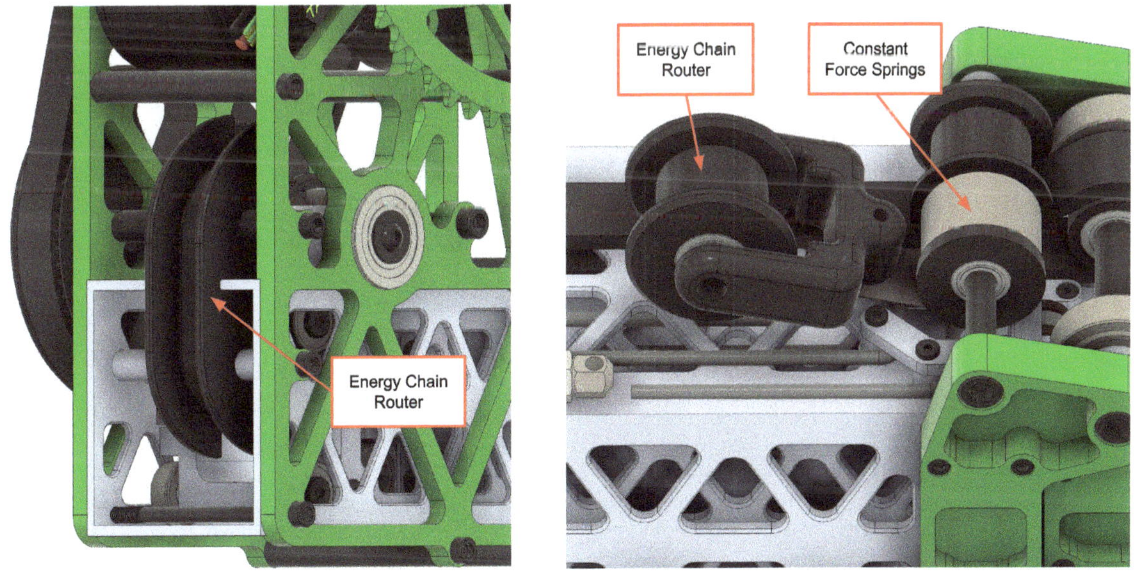


Diag. 9.18: Extendable Arm Ex. 3 Zoomed In View 5; Ref. 33

The chain routers shown in Diagram 9.18 ensure the energy chain follows a specific path through the robot. The energy chain router on the left of Diagram 9.18 is mounted to the static stage of the elevator, whereas the router on the right is mounted to a constant force spring, allowing the energy chain to travel the full length of the static stage of this lift. Analogizing to a telescoping lift, (see Subchapter 8.5 on telescoping lifts): The energy chain acts as the "rope" that pulls on the energy chain router to the left while the constant force spring passively pulls on the energy chain router to the right. There is a resultant small force from this constant force spring on the energy chain, which pulls on the lift to retract. However, this spring is relatively weak as its purpose is to manage the energy chain and not to aid in the retraction of the lift.

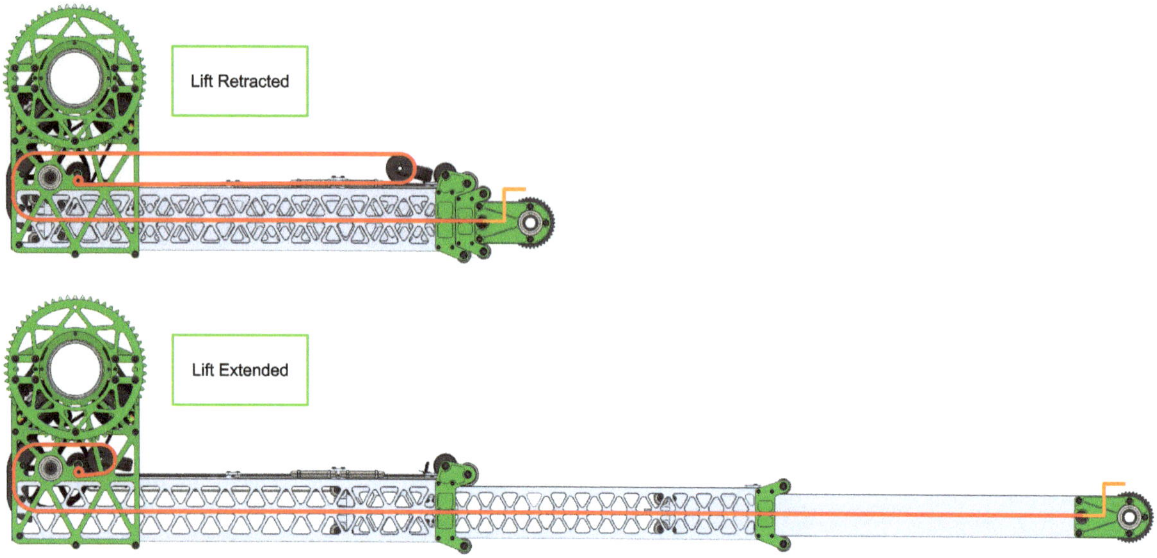


Diag. 9.19: Extendable Arm Ex. 3 Side View 2; Ref. 33

The energy chain route is shown in Diagram 9.19. The small red circle shows the approximate location where the wires enter the lift, and the orange line branching off shows where the wires exit the lift to power the claw of this design.

Standard energy chain is only able to bend in one direction to make the routing of the energy chain more predictable. (Bidirectional energy chains can make sense for certain designs like a turntable, but in the case of a compact lift, such an energy chain may twist and turn in unpredictable ways and get caught in other parts of the mechanism.) Energy chains will always maintain a constant length, but where that length travels changes with the positioning of the mechanism. In the case of this design, as can be seen in Diagram 9.19, the energy chain is only able to curl counter clockwise. In the retracted position, the energy chain travels the length of the first stage 3 times. In the extended position, most of the length of the energy chain runs across the 3 stages of the lift. As all lift stages are similar in length, the two positions show an energy chain route that is the exact same length. The usage of the constant force spring described earlier ensures a smooth transition of the routing of the energy chain between the two shown positions. Running the wiring of the claw in this manner prevents it from getting caught on any part of the lift, thus reducing the risk of damage and ensuring the claw's functionality.

To highlight how versatile an extendable arm design is, Team 2910 adapted the general concept of this design when making a robot for the 2025 game.

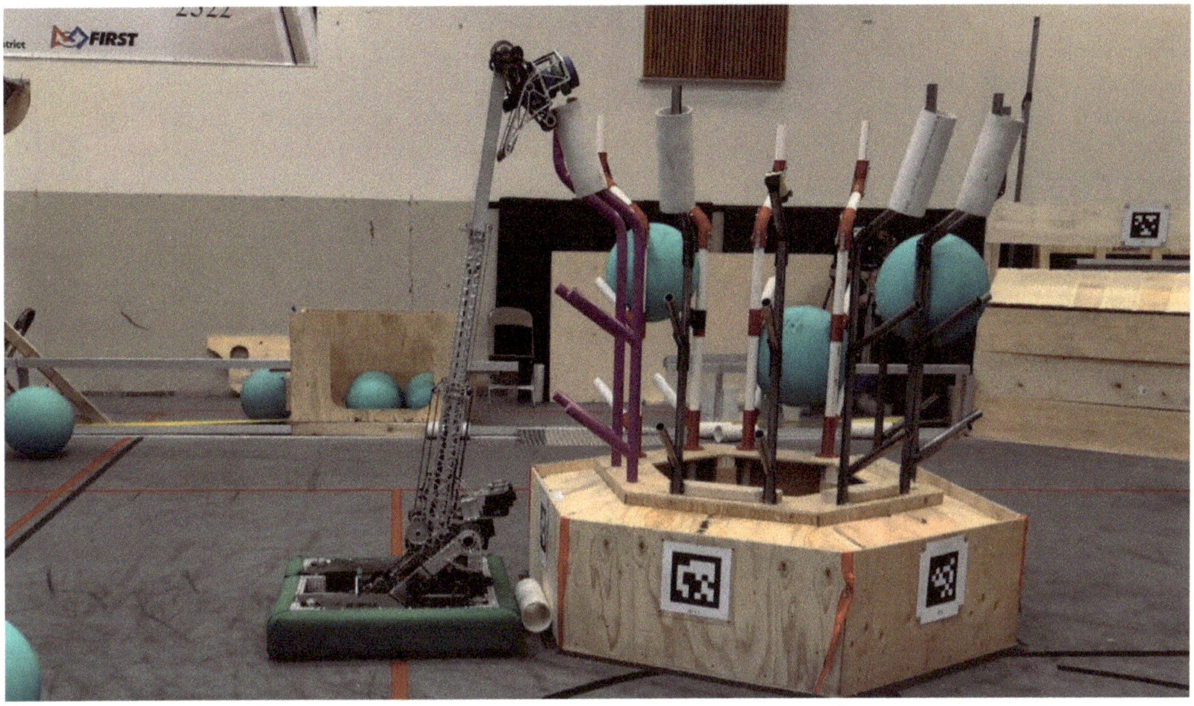

Diag. 9.20: Extendable Arm Ex. 3 Side View; Ref. 34

While the claw design was changed to accommodate the different game pieces and perhaps the length of the stages to give this version of the arm a bit more reach, the general concept of a compact cascade lift on a pivot arm is exactly the same. (To compare the similarities between the 2 designs see https://www.frc-handbook.com/ref/33 and https://www.frc-handbook.com/ref/34.)

## 9.4 Multi Jointed Arms

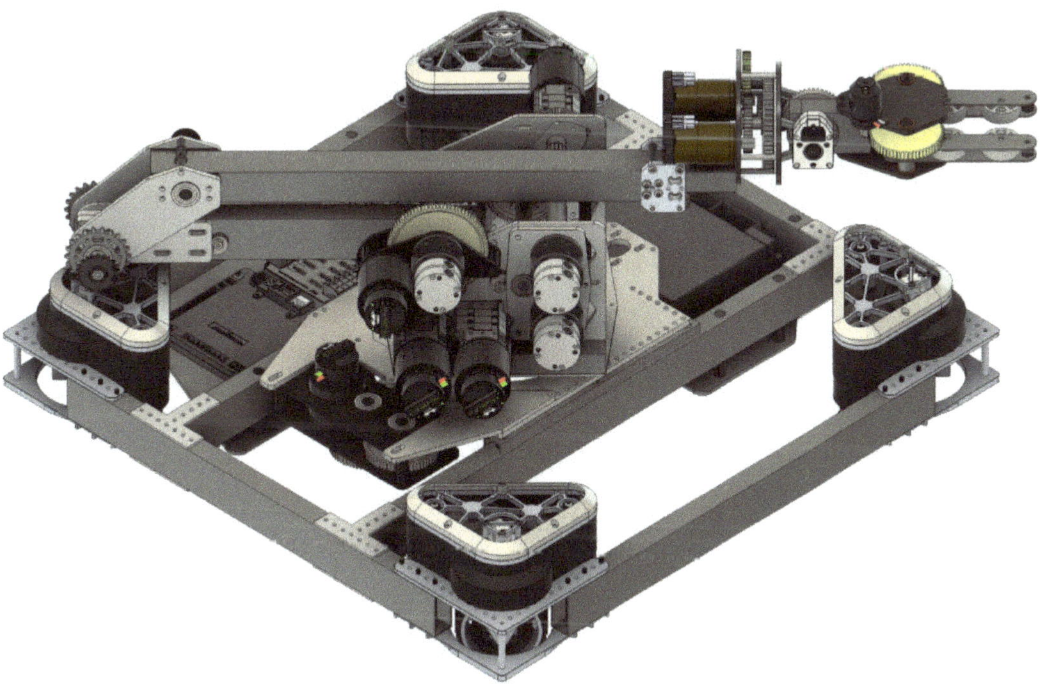

Diag. 9.21: Multi-Jointed Arm Ex. Isometric View; Ref. 35

The example in diagram 9.21 (refer to https://www.frc-handbook.com/ref/35) showcases a complicated multi-axis arm. This was designed by Team 4817 as an off-season project to test the team's capability in designing, building, and programming a complicated arm but for various reasons this project did not get past the CAD stage. Nevertheless, there are some theoretical learnings to point out.

This is a 6 axis arm:

1. The turntable pivot
2. The first pivot arm joint
3. The second pivot arm joint
4. The linear extender (a single stage rope lift) in the second arm
5. The first wrist axis of the claw
6. The second wrist axis of the claw

When designing such a mechanism, relatively high torque gearboxes are required due to the potential for the center of mass to be really far away from the pivots of arms, particularly the first pivot arm. Additionally, the motor would likely strain to hold various positions. As such, pneumatic brakes are incorporated in the design for the 2nd to 4th axis.

To minimize the weight of the arm, gearboxes for the 2nd to 4th axis are driven by a series of chains (and a pulley for the 4th axis) as opposed to placing the gearboxes for each axis beside each pivot/winch.

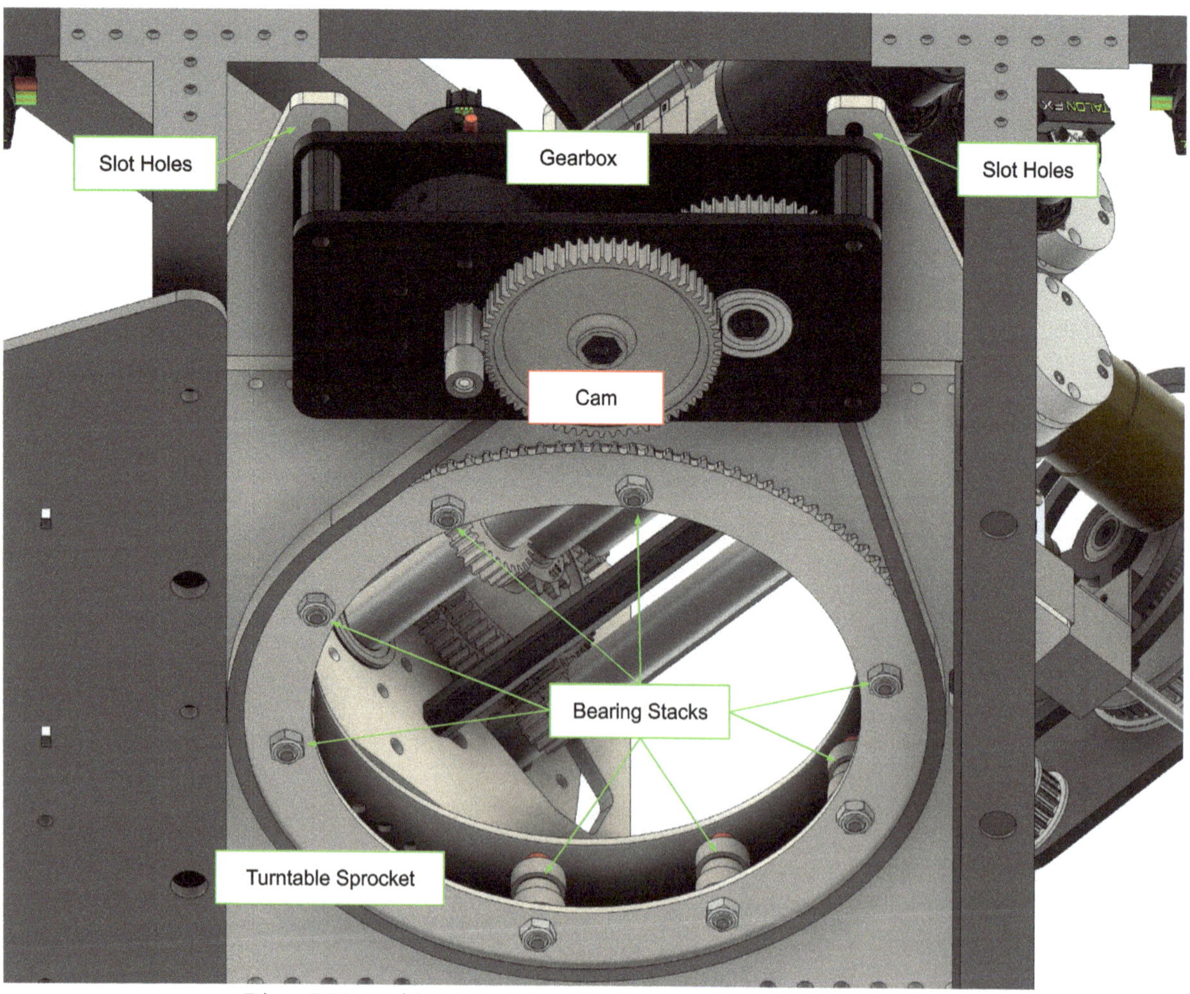


Diag. 9.22: Multi-Jointed Arm Ex. Bottom Slanted View; Ref. 35

This turntable is the first axis, allowing the whole multi-axis arm to rotate relative to the drivebase. A gearbox drives a sprocket that turns a giant sprocket that effectively makes up the turntable. Bearing stacks are used for this turntable design given its large size.

To tension the chain in this system, the entire gearbox is mounted on slot holes. A snail cam positioned where the red box titled "Cam" is, (it can't be seen from this angle in diagram 9.22), pushes on the gearbox system away from the turntable sprocket. This moves the entire gearbox away from the turntable sprocket until the chain is fully tensioned.

The gearbox that drives axis 2-4 is rather complicated. It will be described one axis at a time.

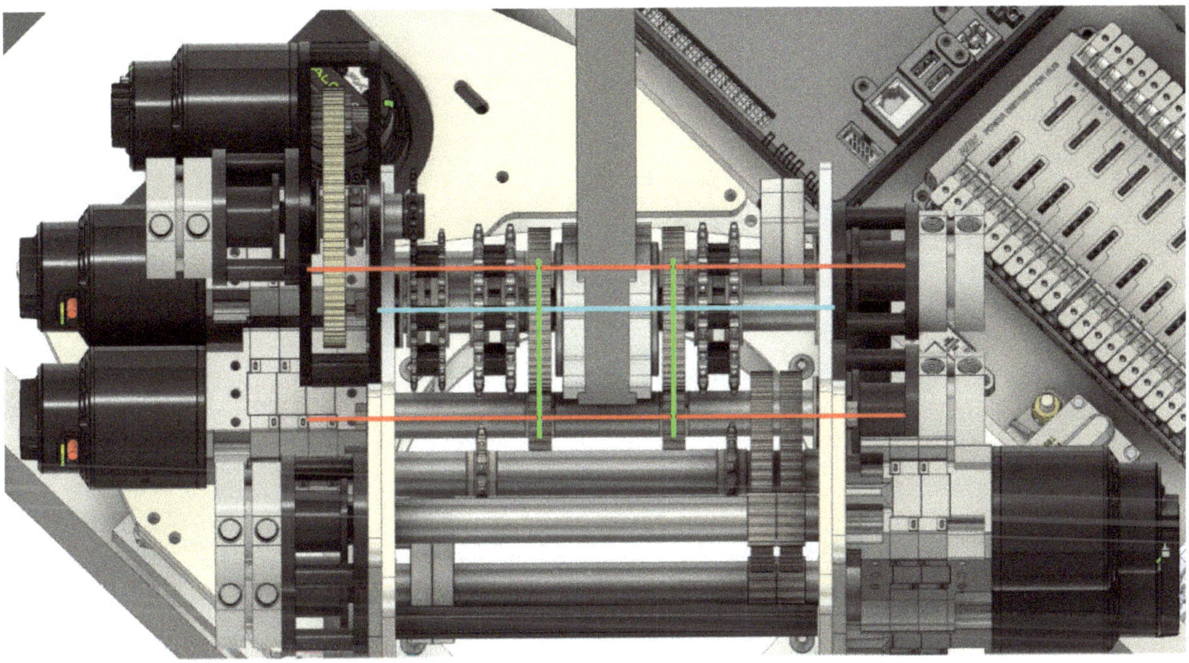

Diag. 9.23: Multi-Jointed Arm Ex. Top Slanted View 1; Ref. 35

This is the second axis, which is also the first pivot arm. 2 motors reduced by MaxPlanetary gearboxes feed into the 2 driving axles, highlighted by red lines in diagram 9.23. At the other end of the driving axles are pneumatic brakes to lock the arm at specific positions. The gears on the driving axles engage with a gear on the primary axle, highlighted by the cyan line. The green lines highlight power transmission for this axis. The driven gear and pivot arm are interlocked to rotate with the axle.

Diag. 9.24: Multi-Jointed Arm Ex. Side View 1; Ref. 35

The side view in diagram 9.24 shows another perspective of the power transmission of the second axis. The red circular arrows highlight the driving axles, while the cyan circular arrows show the movement of the primary axle. The encoder used to track the position of the first pivot arm is also indicated.

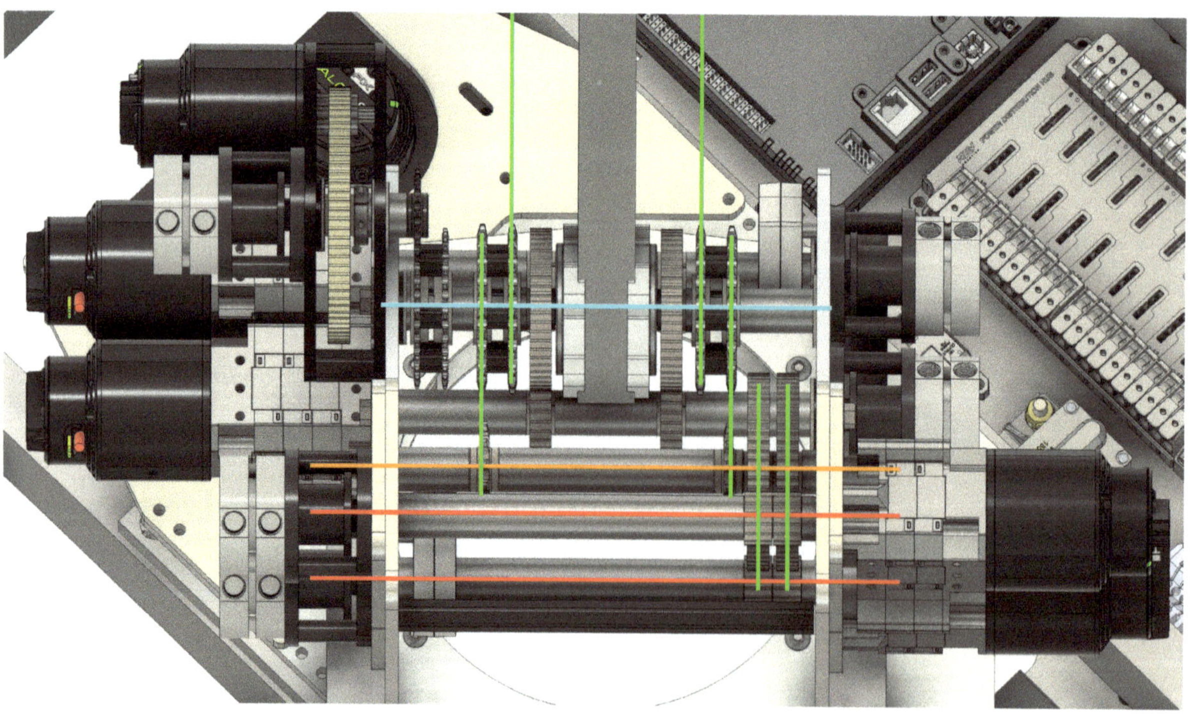

Diag. 9.25: Multi-Jointed Arm Ex. Top Slanted View 2; Ref. 35

Diagram 9.25 portrays the third axis, which is the second pivot arm. Again, the 2 motors are geared down by MaxPlanetary gearboxes, highlighted by red lines. Similarly, at the other end of the driving axles are pneumatic brakes to lock the arm into specific positions. The gears on the driving axles turn gears on a secondary axle, highlighted by the orange line. Then the power transmission feeds into sprockets on the primary axle, highlighted by the cyan line. The sprockets on the primary axle are free spinning to allow the second pivot arm to turn independently of the first pivot arm as the first pivot arm is locked to the rotation of the primary axle. This is done by inserting bearings inside the plate sprockets used for the second pivot arm. The green lines highlight all of the power transmission for this axis.

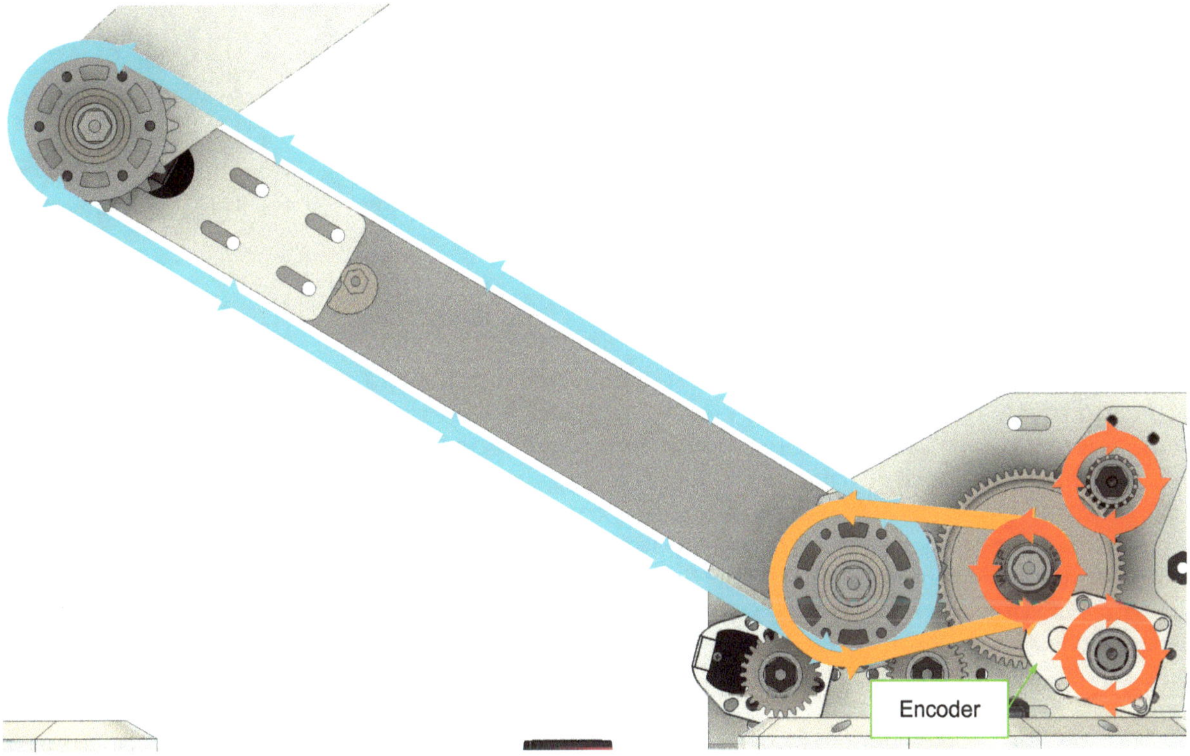


Diag. 9.26: Multi-Jointed Arm Ex. 2 Side View 2; Ref. 35

This side view shows another perspective of the power transmission of the third axis. The red circular arrows show the direction in which the gears move for the driving and secondary axles. The orange arrow set shows the chain wrapped around the secondary axle and the primary axle. And finally, the cyan arrow set shows the chain wrapped around the primary axle and the second arm's pivot axle. Diagram 9.26 also highlights the encoder used to track the position of the second pivot arm.

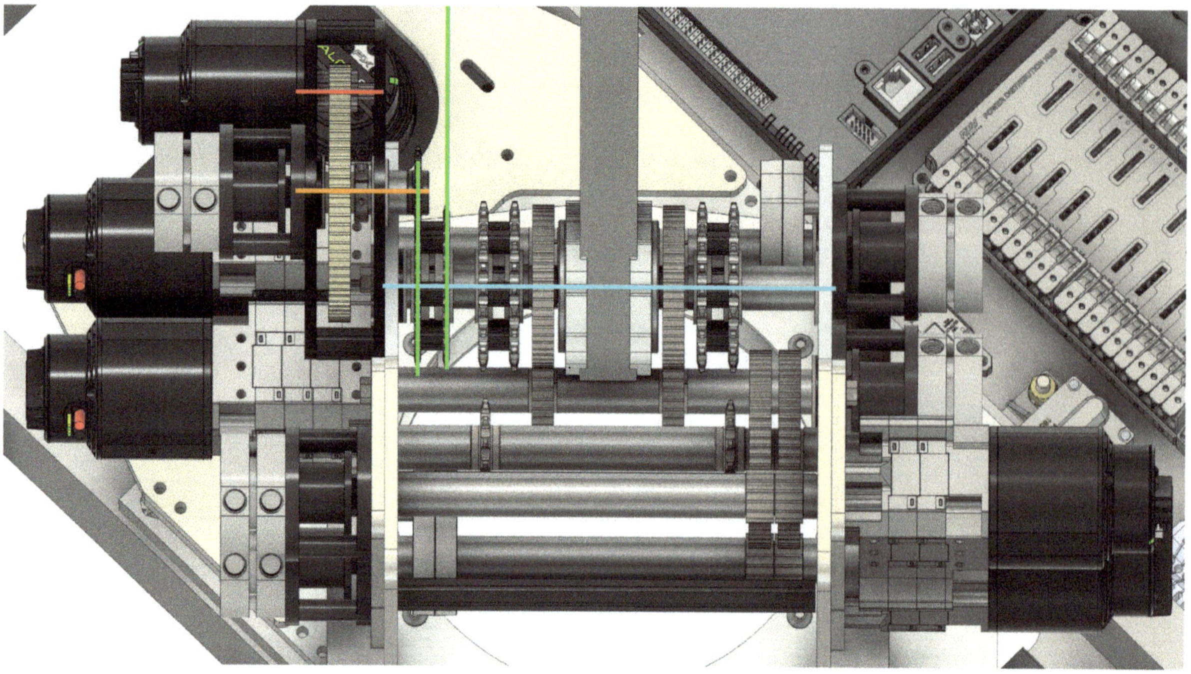

Diag. 9.27: Multi-Jointed Arm Ex. Top Slanted View 3; Ref. 35

Diagram 9.27 shows the fourth axis, which is the linear slide mounted on the second pivot arm. Since a smaller gear reduction is required for the linear slide, it is powered by a single motor without any MaxPlanetary gearboxes. The gear on the driving axle, highlighted in red, turns a gear on a secondary axle, highlighted by the orange line. The pneumatic brake for this mechanism is on the secondary axle. The sprockets on the primary axle are, again, free spinning to allow the linear slide to turn independently of the first pivot arm as the first pivot arm is locked to the rotation of the primary axle. This is done by inserting bearings inside the plate sprockets used for the linear slide. The green lines highlight all of the power transmission for this axis.

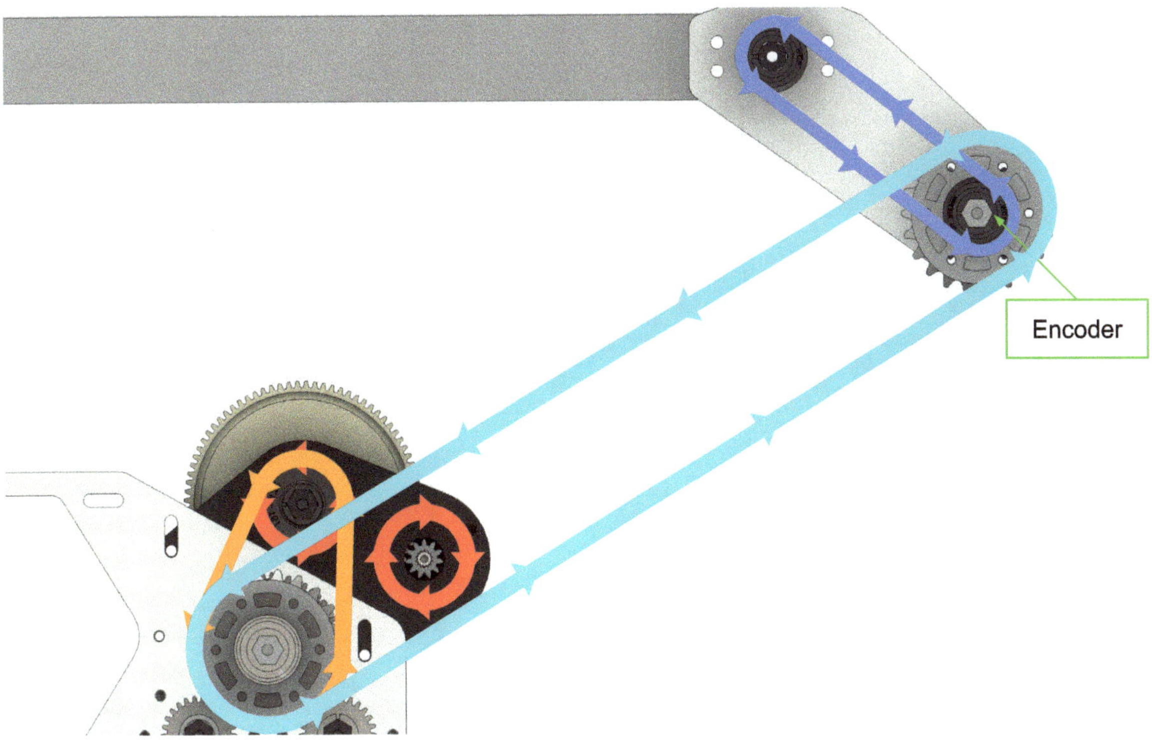


Diag. 9.28: Multi-Jointed Arm Ex. Side View 3; Ref. 35

Diagram 9.28 shows the side view perspective of the power transmission of the fourth axis. The red circular arrows show the gears' rotation in the driving and secondary axles. The orange arrow set shows the chain wrapped around the secondary axle and the primary axle. The cyan arrow set shows the chain wrapped around the primary axle and the second arm's pivot axle. The dark blue arrow set shows the pulley wrapped around the second arm's pivot axle and the axle that drives the winch for the linear slide. The encoder measures the position of the linear slide from the second arm's pivot axle. This is possible because the sprockets that drive the second pivot arm are directly mounted onto the structure and do not turn with the axle.

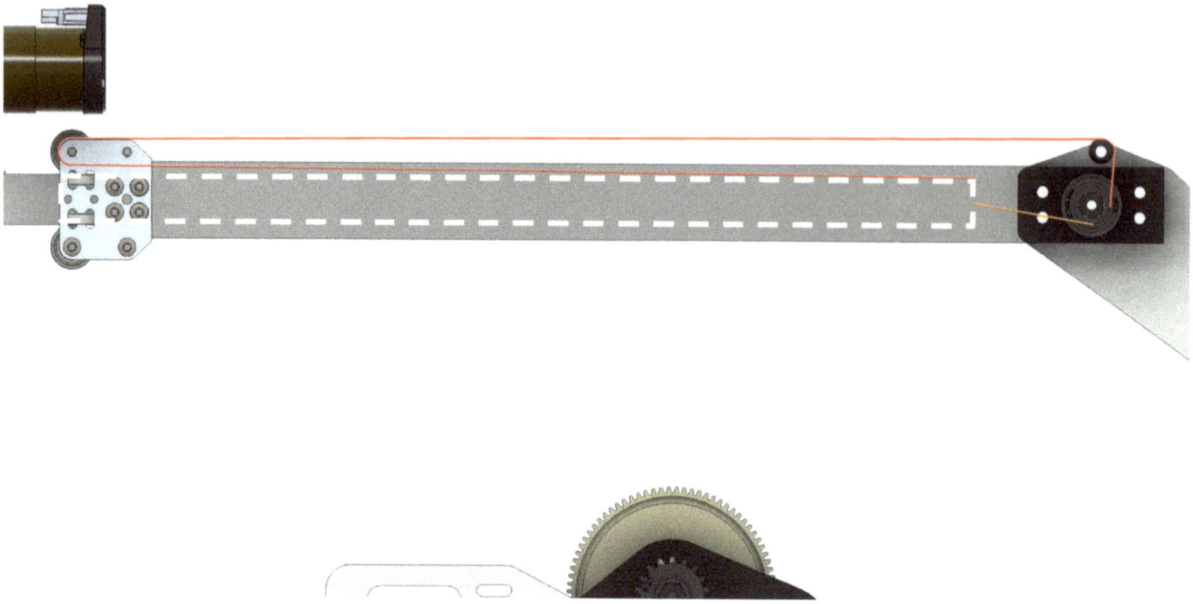

Diag. 9.29: Multi-Jointed Arm Ex. Side View 4; Ref. 35

To expand on how the linear slide works, it is a 2 spool single stage elevator. A winch on the axle, where the final driven pulley is mounted, turns 1 of 2 directions to pull on either the red rope or the orange rope. The hidden stock is outlined by the dotted white lines. Both the red rope and the orange rope are tied roughly around the far right of this stock. By using telescoping lift bearing modules (see Subchapter 8.5 for full description on telescoping lifts) the inner stock is able to move smoothly inside the outer stock. Rotating the winch clockwise reels in the red rope, while letting out the orange rope, resulting in the linear slide extending. Rotating the winch counter clockwise does the reverse.

The final 2 axis of this 6 arm axis acts as a wrist for a claw. See Subchapter 10.5 for more details on wrist designs.

In general, every axle with a sprocket that is not the primary axle is on a structure that is mounted with slot holes so that a cam can tension the chains wrapping around the various sprockets.

## 9.5 Simple Arm & Lift Combinations

Aside from the multi-axis arms mentioned in this chapter so far, there are simpler ways to use arms and lifts in conjunction with each other.

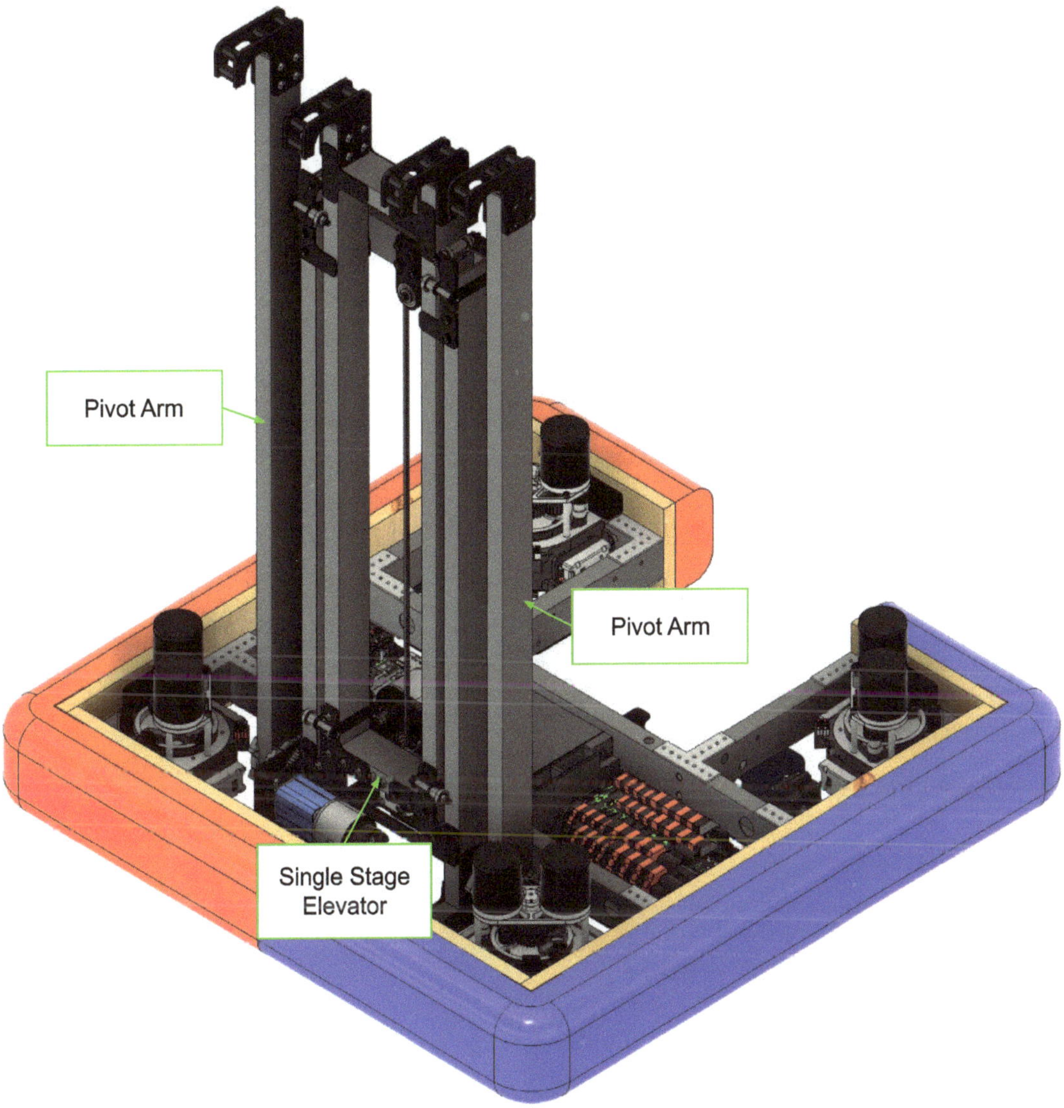


Diag. 9.30: Lift and Arm Combination Ex. Isometric View; Ref. 12

A simple example of a multi-axis mechanism is the combination of a pivot arm and a single stage elevator to reach the highest climb level for the 2022 FRC game that this robot was designed for.

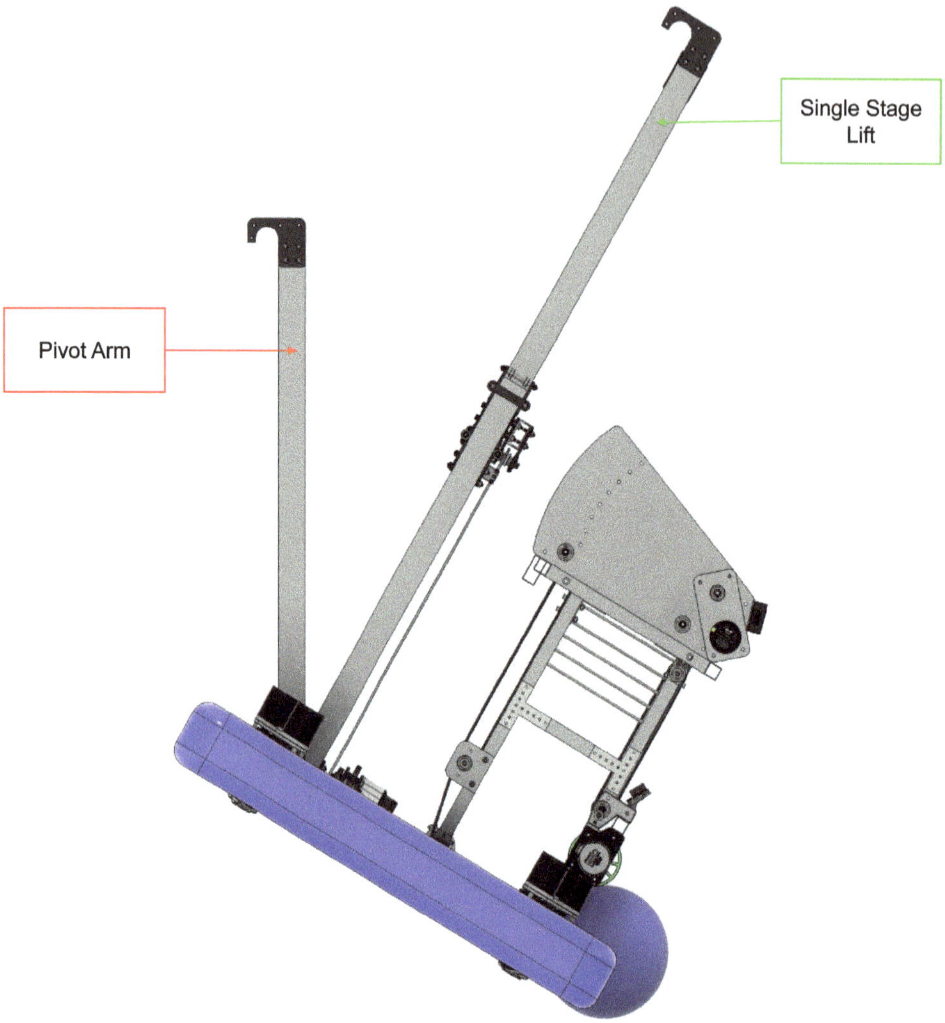


Diag. 9.31: Lift and Arm Combination Ex. Side View; Ref. 12

In order to traverse monkey bars to score more points for the climbing challenge for the 2022 game, this design made use of a single stage lift to reach the second climbing bar. Then the climbing bar would be passed to the pivot arm's claw so that the lift's claw could be used to grip the third climbing bar, and so on. For information on static claws, see Subchapter 10.2. (For more detail on this robot design see https://www.frc-handbook.com/ref/12.)

## 9.6 Summary on Multi-Axis Arms

The more advanced multi-axis arms options shown in this chapter are highly dynamic mechanisms that allow teams to to accomplish tasks not achievable with standard lifts or arms. That being said, it is always better to use a simpler mechanism to accomplish a task if it is possible to do so. Only use the multi-axis arms described in this chapter when necessary.

The foregoing discussion on multi-axis arms is summarised below:

| Arm on Lift |
| --- |
| • First axis is linear; uses a lift mechanism<br>• Second axis is rotational; uses an arm mechanism<br>• Creates an arm that is able to change the position of its pivot in a linear direction<br>• Hard to program if moving both axis together, though program can be simplified to just move one axis at a time |
| **Extendable Arms** |
| • First axis is rotational; uses a pivot arm<br>• Second axis is linear; uses a lift mechanism<br>• Creates a pivot arm that is able to extend and retract<br>• Very hard to program |
| **Multi Jointed Arms** |
| • First axis rotational; uses a pivot arm or a chain bar arm<br>• Second axis is rotational; uses a pivot arm or a chain bar arm<br>• Additional axis can be made to allow for more flexibility<br>    ○ More rotational points can be made using either pivot arms or chain bar arms<br>    ○ Linear slides can be added by using lift mechanisms<br>• Creates a lot of mobility and flexibility in end effector manipulation<br>• Considered to be the hardest mechanism to program (difficulty increases with the addition of axes; effective programming requires understanding of linear algebra) |
| **Turntables** |
| • Turntables can be added to the bottom of any type of arm/lift, even the ones described in Chapters 7 and 8.<br>• Rotational axis around the base of the arm/lift allows for arm/lift to face a direction independent of the drivebase's orientation |

Diag. 9.32: Multi-Axis Arms Summary Table

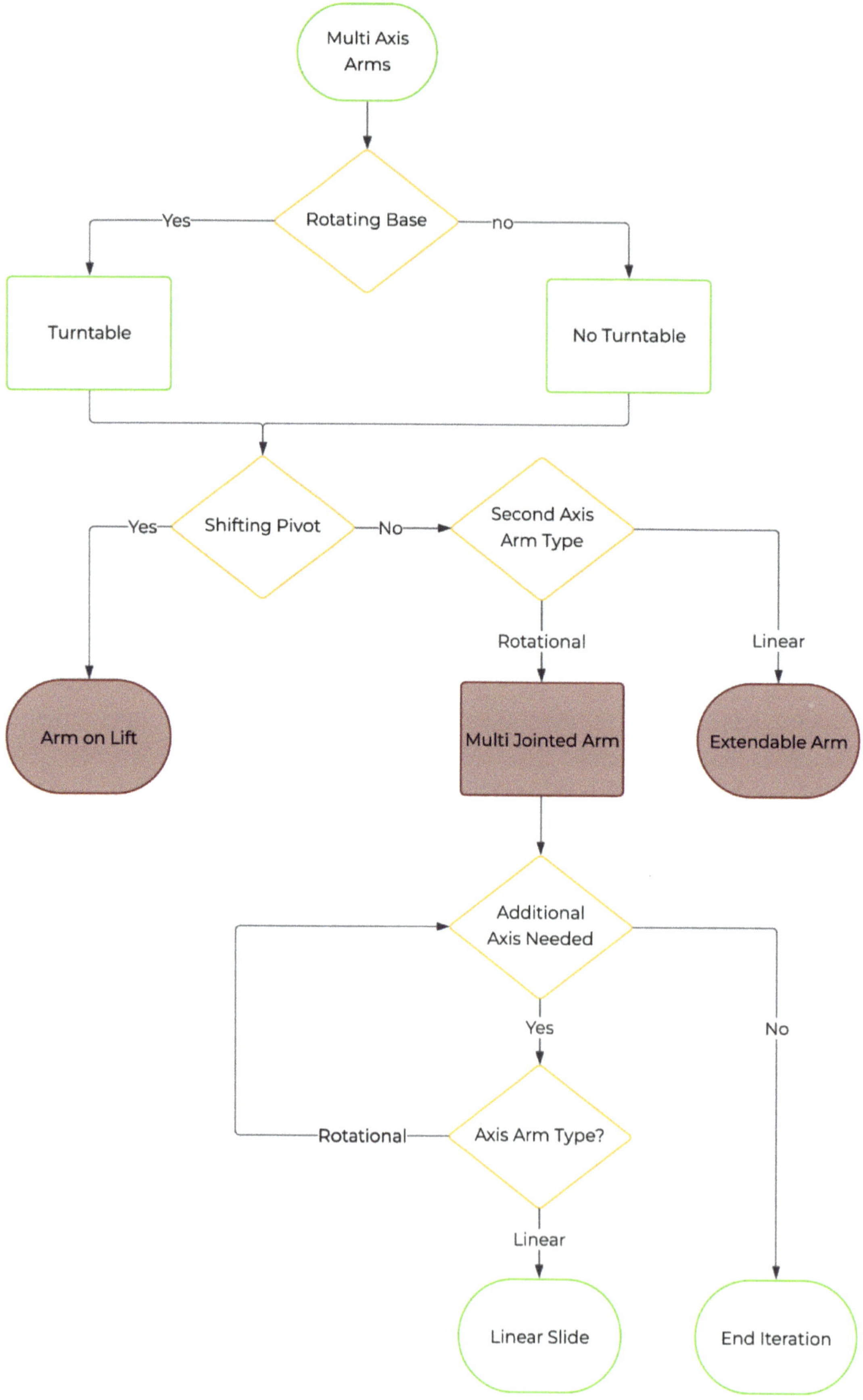


Diag. 9.33: Multi-Axis Arms Flowchart

The flowchart on the left provides a methodology for choosing the type of multi-axis arm needed for a given task. This assumes the designer already has an idea of the required motion—rotational, linear, or a combination—and the number of axes. One way to determine the necessary axes is to act out the arm movements with your own arms and hands, visualizing how the end effector needs to move through space to accomplish the intended task.

Note, technically, it is possible to combine linear and rotational axes in any combination (for example a multi-axis arm with the first axis being rotational, then linear, ending with rotational). However, the difficulty of designing such mechanisms with pivot points that can travel linearly is extreme, considering the mechanics and wiring challenges. The Arm on a Lift example shows the careful consideration required for wiring the pivot arm in that design. Given this, the flowchart recommends using designs that are either a single rotational pivot on a linear axis (an arm on a lift style multi-axis arm), or using designs where the linear axis is the final axis before the end effector. Generally speaking, the kind of motion that such multi-axis arms are able to achieve can usually be accomplished by the recommended versions found in the flowchart. The versions in the flowchart are easier to design than the alternative combinations.

To reiterate, multi-axis arms should only be used when the desired end effector movement cannot be accomplished by a standard arm (see Chapter 7) or lift mechanism (see Chapter 8). Using simpler mechanisms whenever possible is always best.

# 10. Claws

## 10.1 Introduction

Claws are mechanisms designed to grab and release game objects. Some claw designs can resemble intakes since intakes can also spin in reverse to eject game elements. The distinction lies in function: if the mechanism's role is to bring the game object to another part of the robot, it is considered an intake. If it serves as an end effector and is the only mechanism interacting directly with the game object, it is classified as a claw.

Claws are often mounted at the end of an arm or elevator. This is because a claw is usually needed to hold a game object while the arm or elevator positions it at a specific elevation to score points in the FRC game. However, a claw can also be useful when mounted statically to a tower or even directly to the robot's drivebase.

## 10.2 Static Claws

The static claw is the most basic of all claws. Static claws do not require a motor, hence the word static. Static claws utilize the motion of the arm or lift mechanism to which they are attached to grab objects. Most static claws are simple hook-like structures mounted at the end of a lift, typically used for climbing onto a bar or similar structure.

Here is the simplest example of a static claw:

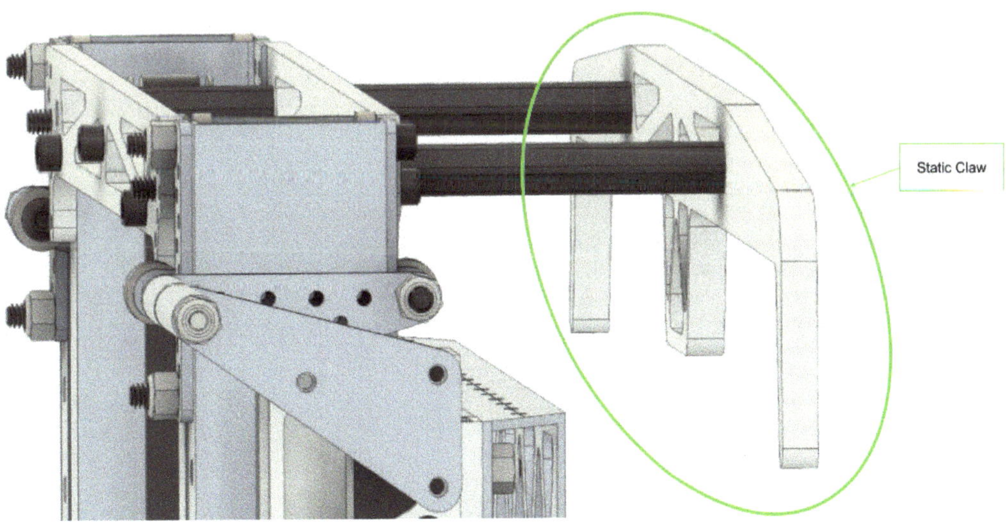


Diag. 10.1: Static Claw Ex. 1 Isometric View; Ref. 28

This static claw design is by Team 2910. The claw is mounted at the top of the continuous spooled lift of Reference 28. (Refer to https://www.frc-handbook.com/ref/28 for more information on the game and robot.)

There are slightly more complicated static claw designs that can interact with game elements in multiple ways.

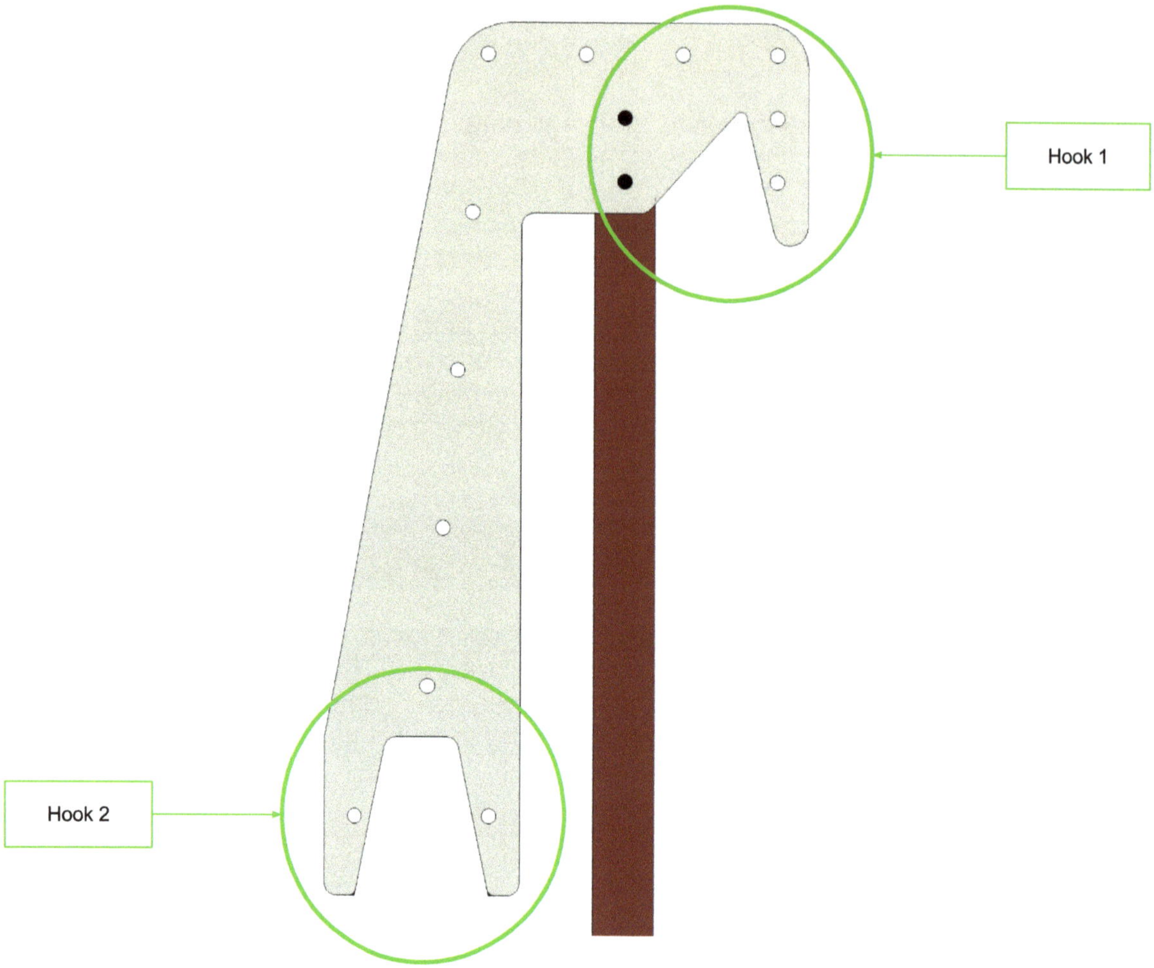


Diag. 10.2: Static Claw Ex. 2 Side View; Ref. 8

This is a static claw design that has 2 different hooks. The design enables the robot to climb the chain to 2 different heights for the 2024 game. (For more information see https://www.frc-handbook.com/ref/8.) One height allows the robot to make a quick climb, while the other enables the robot to potentially score bonus points through a game element called the trap, but would take a longer time to deploy.

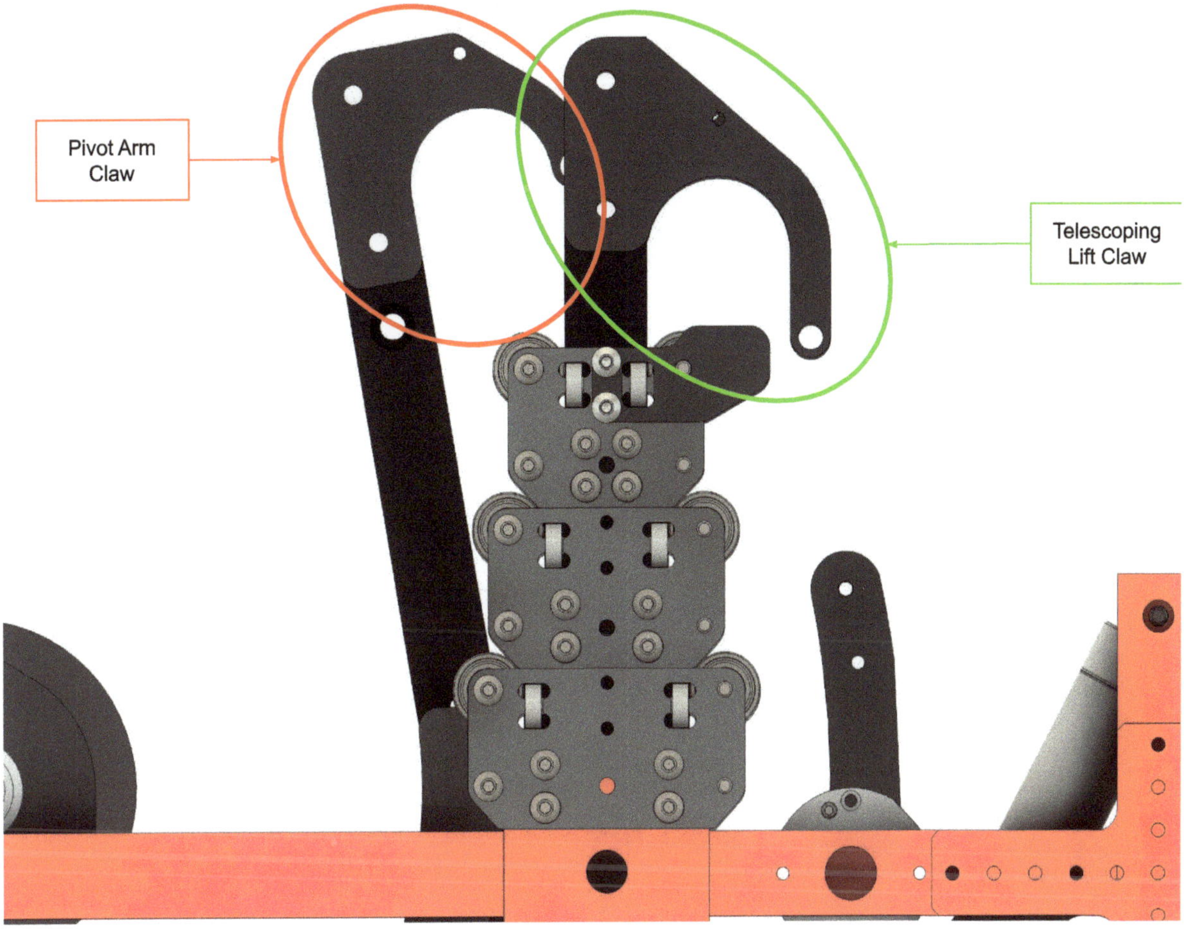


Diag. 10.3: Static Claw Ex. 3 Side View; Ref. 10

The design in diagram 10.3 makes use of 2 static claws: 1 at the end of the telescoping lift, circled in green, and the other at the end of a pivot arm, circled in red. First, the telescoping lift would be used to pull the robot up to the climbing bar. Then, the pivot arm would be used to lock on to the climbing bar by rotating to align with the telescoping lift's claw. This would then allow for the strain of hanging on to the climbing mechanism to be transferred from the telescoping lift to a more stable pivot arm. See https://www.frc-handbook.com/ref/10 for more information.

As a general note, when designing static claws, the claw should be mounted to the end of the lift or arm securely (meaning multiple mounting points) and the outline of the claw should closely follow the profile of the object that the claw will hook onto.

## 10.3 Gripper Claws

### Standard Gripper Claw

A gripper claw makes use of either a linear slide or pivot motion to grab and maintain a grip on game objects. Since some gripper claws make use of pivots, see Subchapter 7.2 on Pivot Arms. All the general information there on pivot arms is about the base design feature that gripper claws make use of.

Refer to https://www.frc-handbook.com/ref/13 for an example of a gripper claw from Team 118.

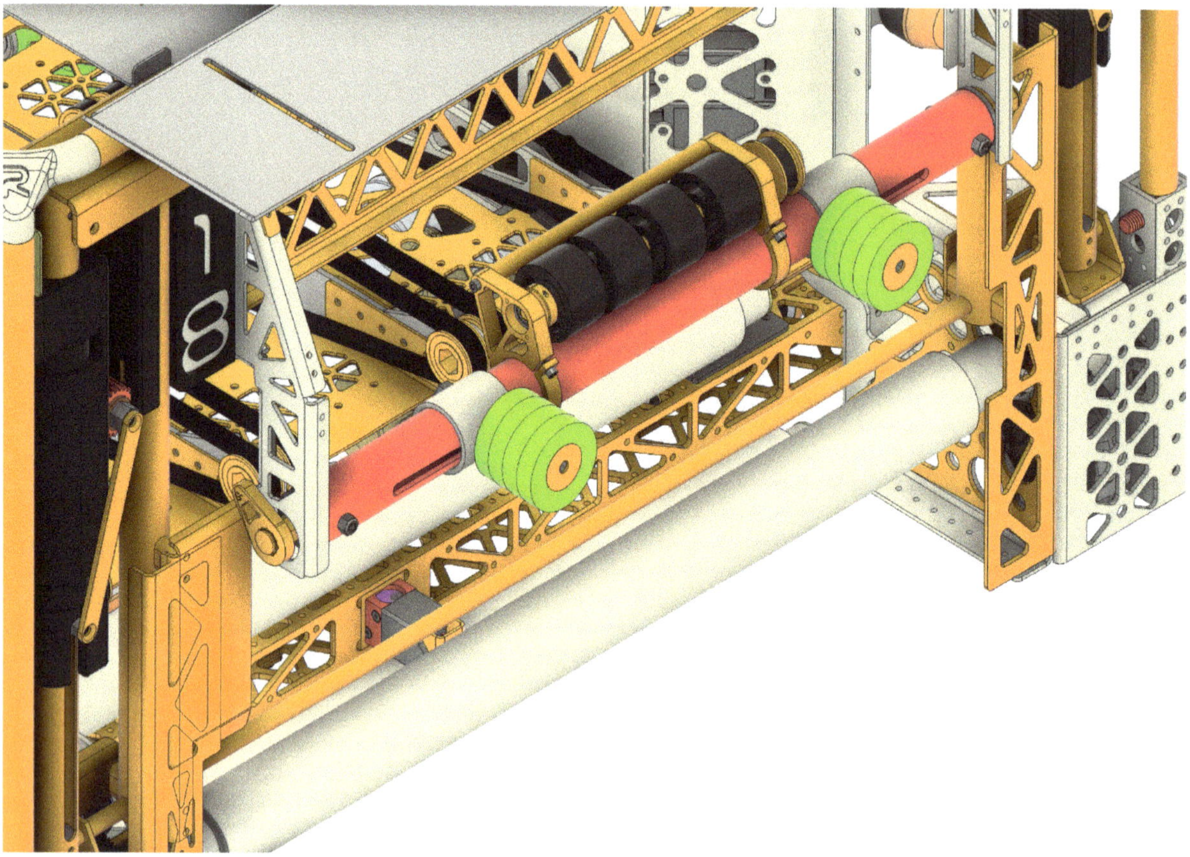

Diag. 10.4: Gripper Claw Ex. 1 Isometric View; Ref. 13

This design features a hidden pneumatic to move the 2 green prongs either closer together or further apart to grab or let go of the gear game object for the 2017 FRC game. This is an example of a linear slide type of gripper claw.

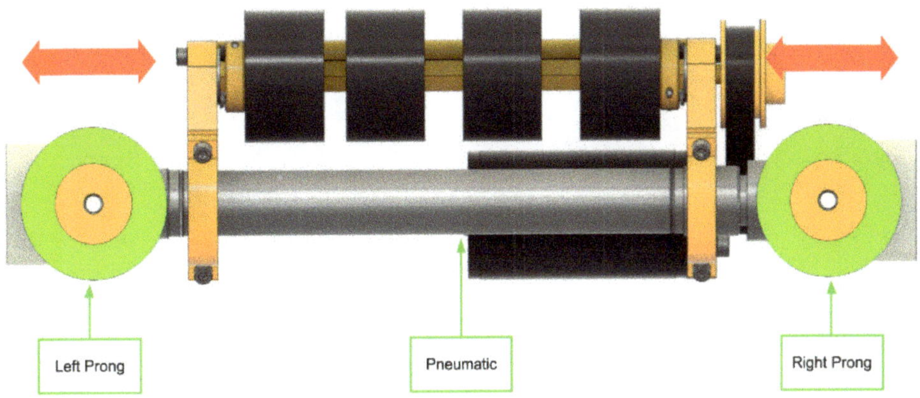


Diag. 10.5: Gripper Claw Ex. 1 Front View; Ref. 13

This view shows the pneumatic and highlights the prongs that move. (The supporting pipe structure was removed.)

Sometimes claws work with intake mechanisms. In these situations, an intake is used to feed the game object to the claw, usually by positioning or aligning the game object in some specific way that makes it easier for the claw to grab the game object. The example in diagram 10.6 and diagram 10.7 is designed by Team 971.

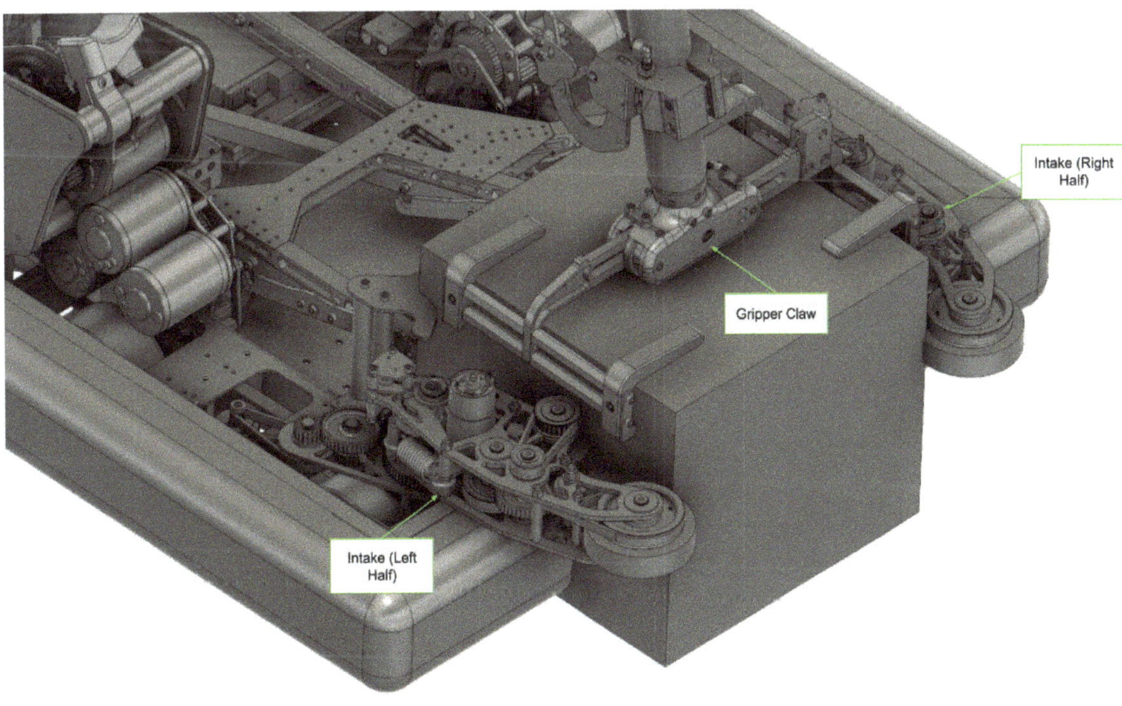


Diag. 10.6: Gripper Claw Ex. 2 Isometric View; Ref. 36

An overbumper intake makes sure that cubes are aligned in a specific way so that this robot's gripper claw can effectively grab onto the cube. (Refer to https://www.frc-handbook.com/ref/36 for more information on the game and full robot.)

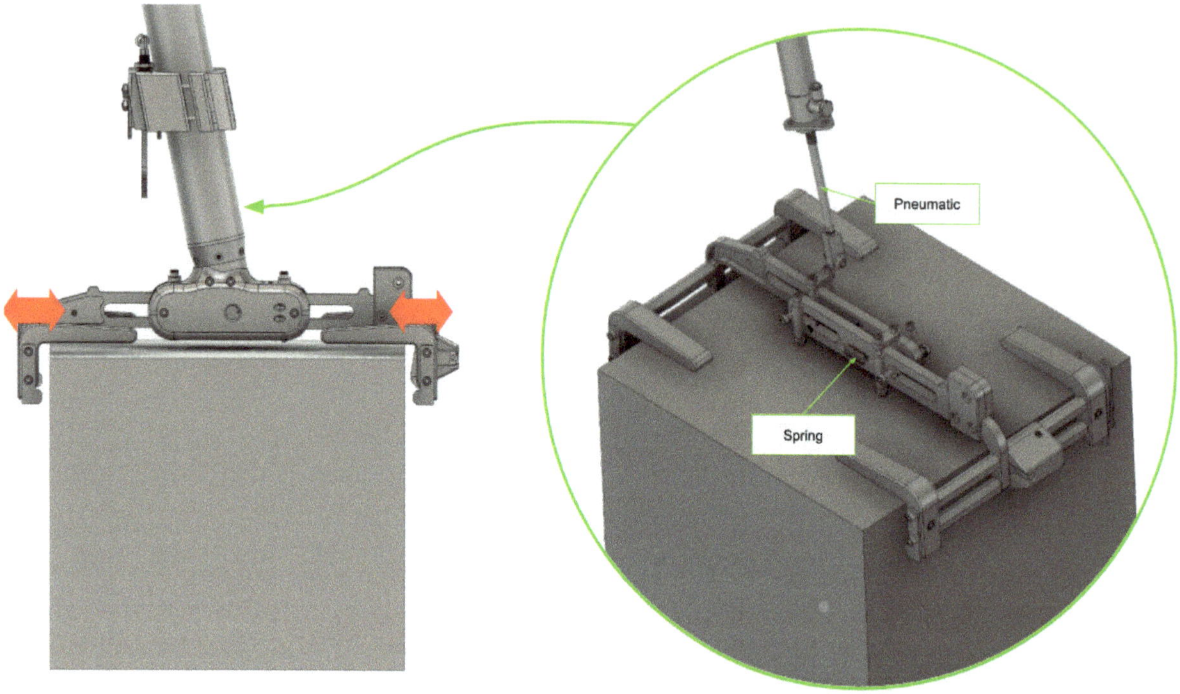


Diag. 10.7: Gripper Claw Ex. 2 Front View; Ref. 36

Once the cube is in place, this gripper claw works through a combination of pneumatics and springs. Springs passively pull the 2 ends of the gripper claw together. The pneumatic when retracted allows the claw to grab onto the cube. The pneumatic when extended effectively drives a wedge to push back against the springs, opening the gripper claw.

In the next example, the gripper claw uses a motor to rotate two sides of the claw for clamping action.

Diag. 10.8: Gripper Claw Ex. 3 Isometric View; Ref. 35

See https://www.frc-handbook.com/ref/35 for more information on the project and full robot.

Diag. 10.9: Gripper Claw Ex. 3 Top View; Ref. 35

In diagram 10.9, a motor drives a small pinion under it that turns the 2 larger yellow gears in opposite directions. This rotates the arms of the gripper claw to either open or close.

Even though there are wheels on this claw, this is not a roller claw since the wheels are statically mounted to the structure of the claw. However, it is possible to combine a gripper claw and a roller claw design together as will be discussed in Subchapter 10.4.

### Suction Gripper Claw

There is yet another type of gripper claw that does not rely on a pivot or sliding structure to grip onto the game object. Instead, these gripper claws use suction.

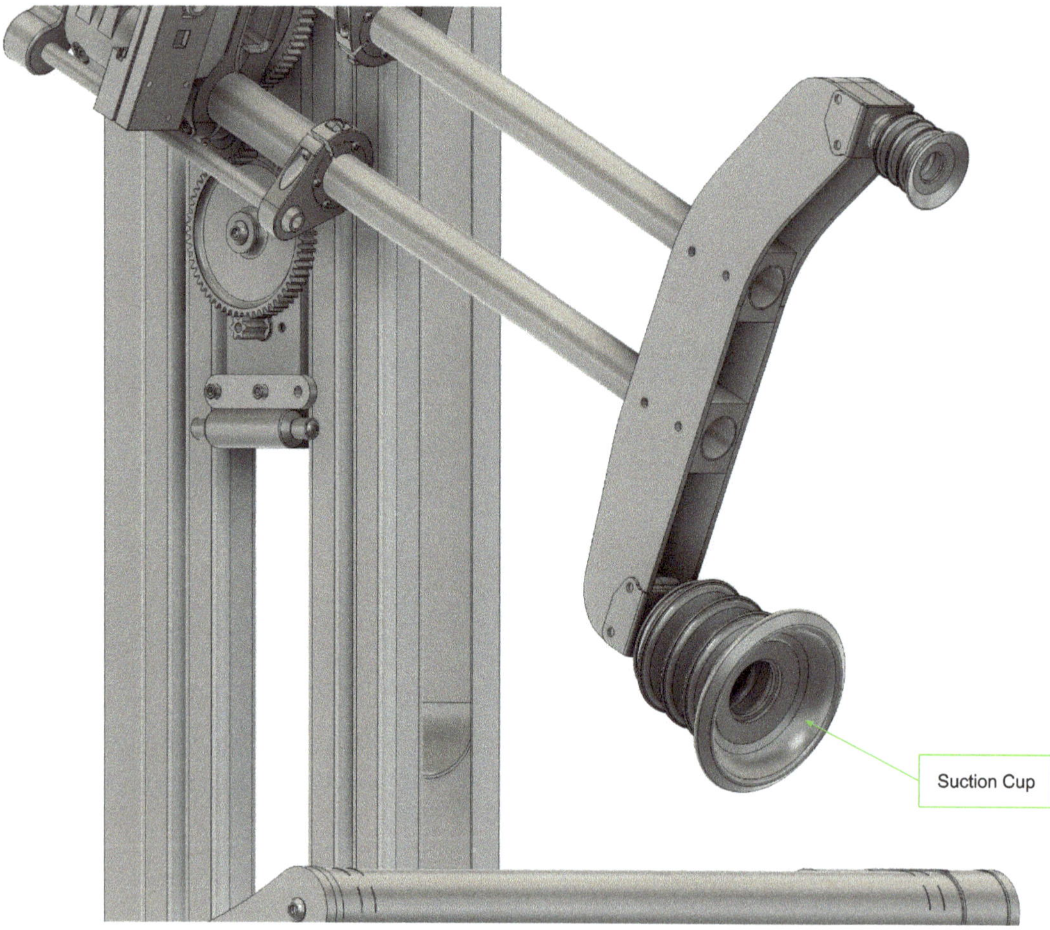


Diag. 10.10: Suction Gripper Claw Ex. 1 Front-Side View; Ref. 37

Team 971 designed the unconventional suction gripper claw example in diagram 10.10. (For more information on the game and full robot see https://www.frc-handbook.com/ref/37.) By treating the suction cup as a pneumatic, when air is pulled out of the suction cup, the gripper claw is able to bind onto a game object. When air is forced back into the suction cup, the gripper claw releases the game object.

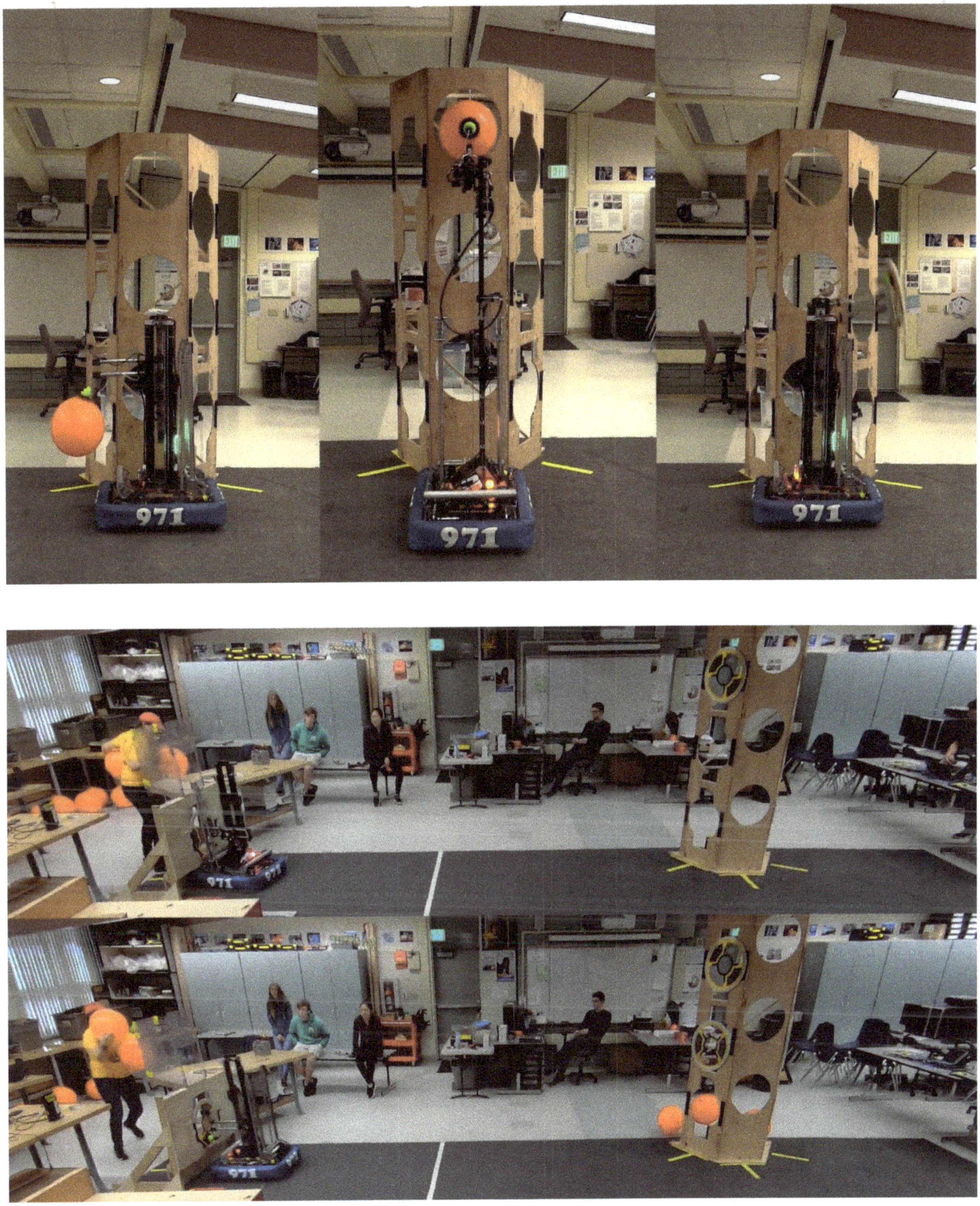


Diag. 10.11: Suction Gripper Claw Ex. 1 Multi View; Ref. 37

This gripper claw is able to grab onto 2 very different game objects: a ball and a flat plate. Multi-tool claws refers to any type of claw that is able to grab multiple types of game objects. Since this claw is able to interact with 2 different game objects, this is an example of a multi-tool claw.

## 10.4 Roller Claws

Roller claws are claws that somewhat resemble intake mechanisms since they use rollers to grab game objects.

While some roller claws will use the same kind of long thin tubing that is commonly found in intake designs, it is more common for a roller claw to use low durometer wheels. Such claws are easier to use, since the stretchy wheels give the driver some tolerance for driving control error. In other words, the driver does not have to align with the game object as accurately because a good roller claw design will be able to grab onto a game object with some variance in its positioning. This general idea also applies to intakes, (intakes should be designed to accept game objects easily). A good roller claw design typically incorporates large rubbery wheels.

This first example makes use of a combination of wheels to effectively grab two different game objects. See https://www.frc-handbook.com/ref/38 for more information on the game and full robot.

Diag. 10.12: Roller Claw Ex. 1 Multi View; Ref. 38

This is an example of a multi-tool roller claw that is able to grab both the cone and the cube for the 2023 FRC game.

Diag. 10.13: Roller Claw Ex. 1 Power Transmission; Ref. 38

The roller claw makes use of a MaxPlanetary gearbox to reduce the gear ratio. Then with a series of 3 gears hidden behind the plate that the gearbox is mounted on, 2 pulleys are driven in opposite directions, outlined by the red arrow sets. These pulleys spin the larger wheels at the bottom of the claw that effectively grab the cube as the smaller wheels up top grab the cone.

By using wheels of varying sizes positioned strategically, this claw functions as a multi-tool claw mechanism. The wheels are spun to either intake or outtake regardless of the type of game object the robot is interacting with.

Refer to https://www.frc-handbook.com/ref/34 for another implementation of a roller claw, by Team 2910.

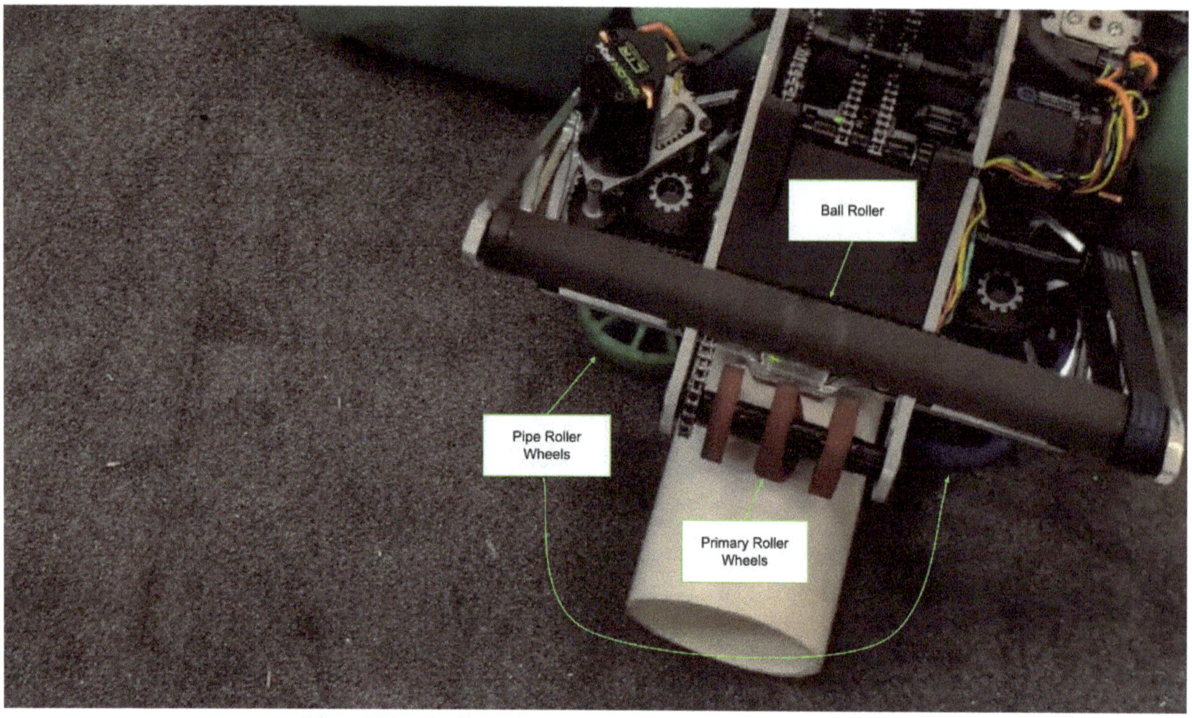


Diag. 10.14: Roller Claw Ex. 2 Zoomed In View; Ref. 34

This multi-tool roller claw provides 2 pathways around the primary roller wheels, as seen in diagram 10.14. When intaking a pipe, the primary roller wheels assist the pipe roller wheels by spinning downward. When the claw is dealing with a ball, the primary roller wheels spin up, while the ball roller spins down to grab the ball. The rollers all spin in the opposite directions to eject the game object.

In reality this claw is one mechanism, and not 2 different roller claws attached near each other. The power transmission of both the top half and bottom half of this claw can be linked because only one side of the claw is being used at any given time. In other words, it does not matter that the ball roller is spinning any which direction while the primary roller wheels assist in grabbing a pipe because the claw is not attempting to grab a ball at the same time. This logic also applies in the other situation when intaking a ball and not a pipe.

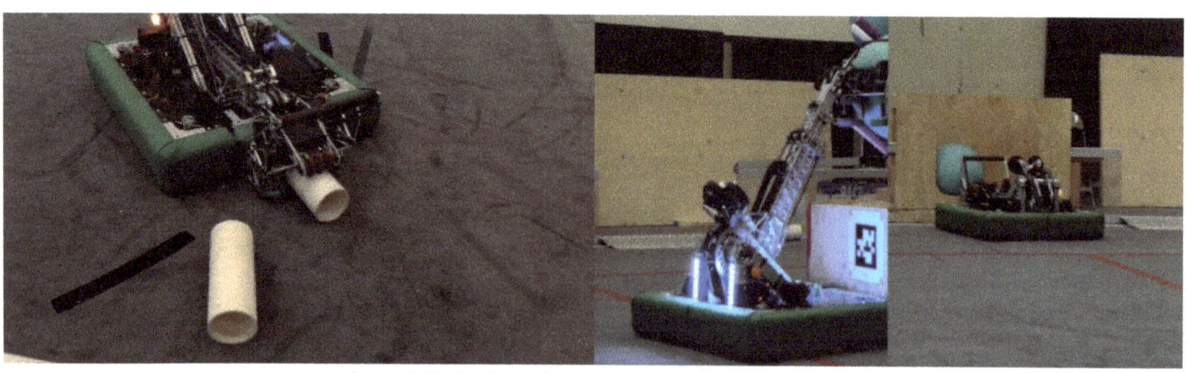

Diag. 10.15: Roller Claw Ex. 2 Multi View; Ref. 34

All of this comes together so that the claw is effective at dealing with both balls and pipes.

While the examples provided of roller claws are all multi-tool claws, to be clear, this does not have to be the case. A claw that uses rollers is perfectly suitable for dealing with a single game object. In general, multi-tool roller claws are easier to make than multi-tool gripper claws.

It is also possible to have both the gripping motion and rolling motion for a claw design. While this may be excessive for certain game objects, here is an example that demonstrates how such a mechanism can be implemented:

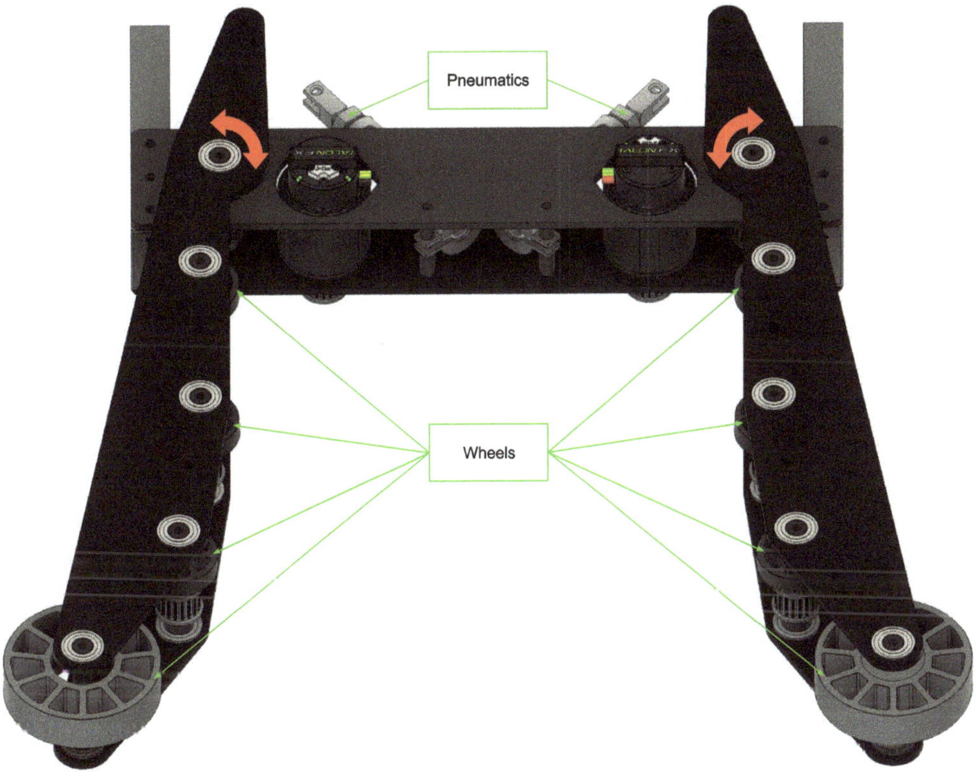


Diag. 10.16: Roller + Gripper Claw Ex. 1 Front Top View; Ref. 27

This claw makes use of pneumatics to rotate and grip the cube and incorporates a series of pulleys to assist in pulling the cube into the robot and shooting it onto the seesaws. See https://www.frc-handbook.com/ref/27 for more information on the game and full robot.

This affords the claw a lot of flexibility in how the claw contacts the cube. The team designed a gripping and rolling mechanism to deal with the cube entering the claw at a slant. If the claw is only a roller claw, it will not be able to open up to facilitate a realignment of the cube against the roller's wheels.

The bottom 2 wheels are set at a very wide position to make it easier for the driver to align the cubes and to grab them with the claw.

## 10.5 Wrist Designs

There are various ways to attach a claw to an arm or elevator. The obvious way is to statically connect the claw with a plate or other simple structure. Almost all of the claw examples shown so far are attached to an arm or elevator in this way.

However, sometimes additional maneuverability is required. Here are some examples of claws attached with a wrist joint.

In this first example, the wrist has one axis.

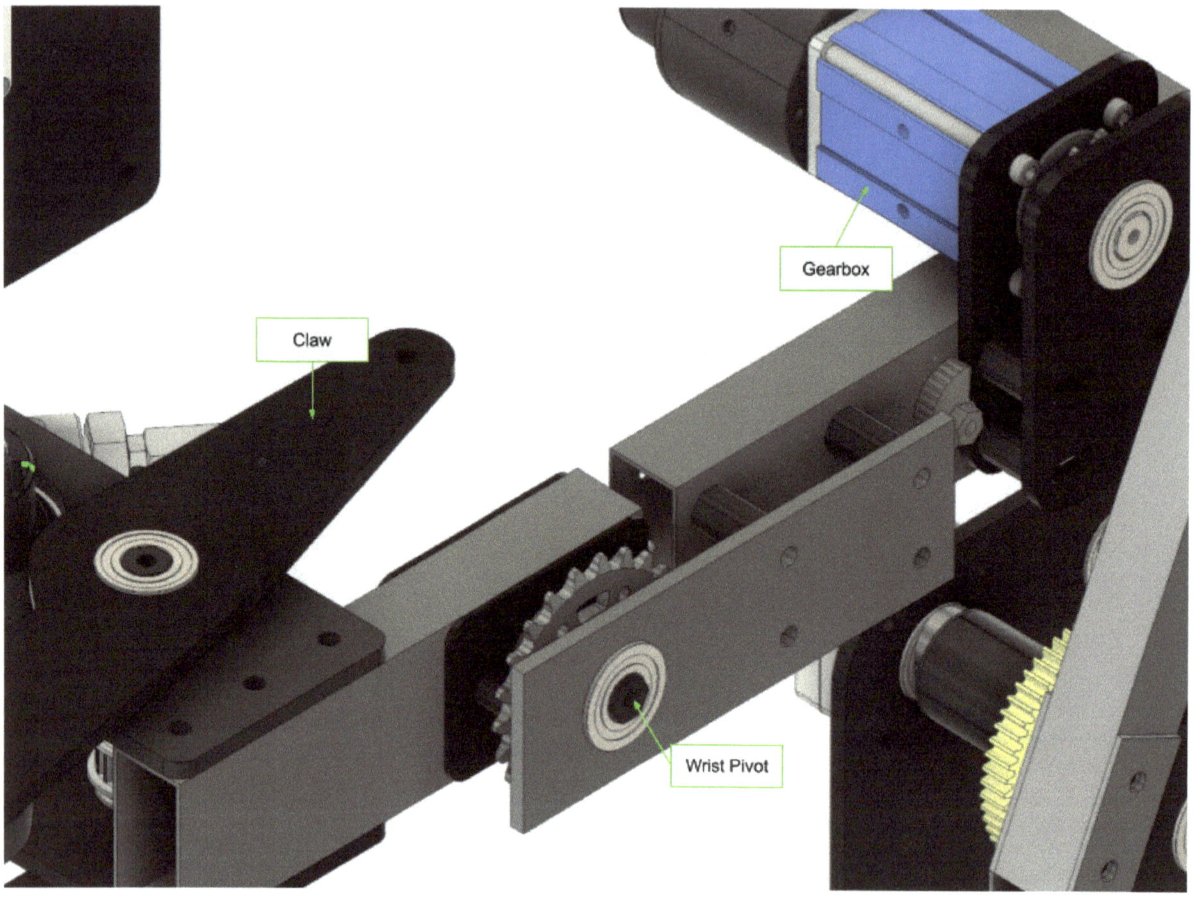


Diag. 10.17: Wrist Ex. 1 Isometric View; Ref. 27

From this view the claw, wrist pivot, and gearbox can be seen and are labeled as such. The wrist pivot is connected to the gearbox by an unpictured chain. See https://www.frc-handbook.com/ref/27 for more information on the game and full robot.

Diag. 10.18: Wrist Ex. 1 Multi View; Ref. 27

Not only does the wrist allow this claw to start within sizing restrictions and then extend beyond frame perimeter to be more effective at collecting cubes, the wrist also allows the claw to rotate to a better release position.

In this next example, the wrist design features 2 pivots.

Diag. 10.19: Wrist Ex. 2 Isometric View; Ref. 35

This design takes inspiration from a common module that has 2 pivots, a swerve drive. The wrist must be able to rotate to multiple positions to handle a soda can effectively. See https://www.frc-handbook.com/ref/35 for more information on the project and full robot.

In essence, the design functions similarly to a swerve drive module. See Subchapter 2.8 on how a swerve drive works. The main difference is that the gearing of the axis that would normally be a wheel is reduced significantly, prioritizing precision control over speed. By replacing a wheel with a piece of stock that the claw is then mounted to, the claw has 2 axis of rotation. One axis allows for a full 360° of rotation, while the other axis is limited to roughly 180° of rotation.

The fact that both motors are on top of the module instead of a rotating axis makes wire management significantly easier.

## 10.6 Summary on Claws

Below is a summary on various claw designs:

<table>
<tr><th colspan="2">Static Claws</th></tr>
<tr><td colspan="2"><ul><li>No moving parts, no motor</li><li>Movement is caused by the manipulator the static claw is attached to (arm/lift/multi-axis arm)</li></ul></td></tr>
<tr><th colspan="2">Gripper Claws</th></tr>
<tr><td>Standard Gripper Claws</td><td><ul><li>Claw performs a gripping motion to grab game objects with prongs</li><li>Gripping motion of prongs is powered linearly with pneumatics or rotationally with gearboxes and motors</li></ul></td></tr>
<tr><td>Suction Gripper Claw</td><td><ul><li>Uses pneumatics and suction cups to bind game object to the claw</li></ul></td></tr>
<tr><th colspan="2">Roller Claws</th></tr>
<tr><td colspan="2"><ul><li>Resembles intakes as claw uses spinning rollers to move game objects in and out of mechanism</li><li>Unlike intakes, which pull game objects in to feed other mechanisms, roller claws are meant to both intake and eject objects</li></ul></td></tr>
<tr><th colspan="2">Wrists</th></tr>
<tr><td colspan="2"><ul><li>Create 1, 2, or 3 axis of rotation</li><li>Axes created by very short "pivot arms" and/or turntable structures</li></ul></td></tr>
</table>

Diag. 10.20: Claws Summary Table

Note that claws as end effectors can be attached to arms (Chapter 7), lifts (Chapter 8) and multi-axis arms (Chapter 9). The flowchart on the right serves as a guide in specifying claws.

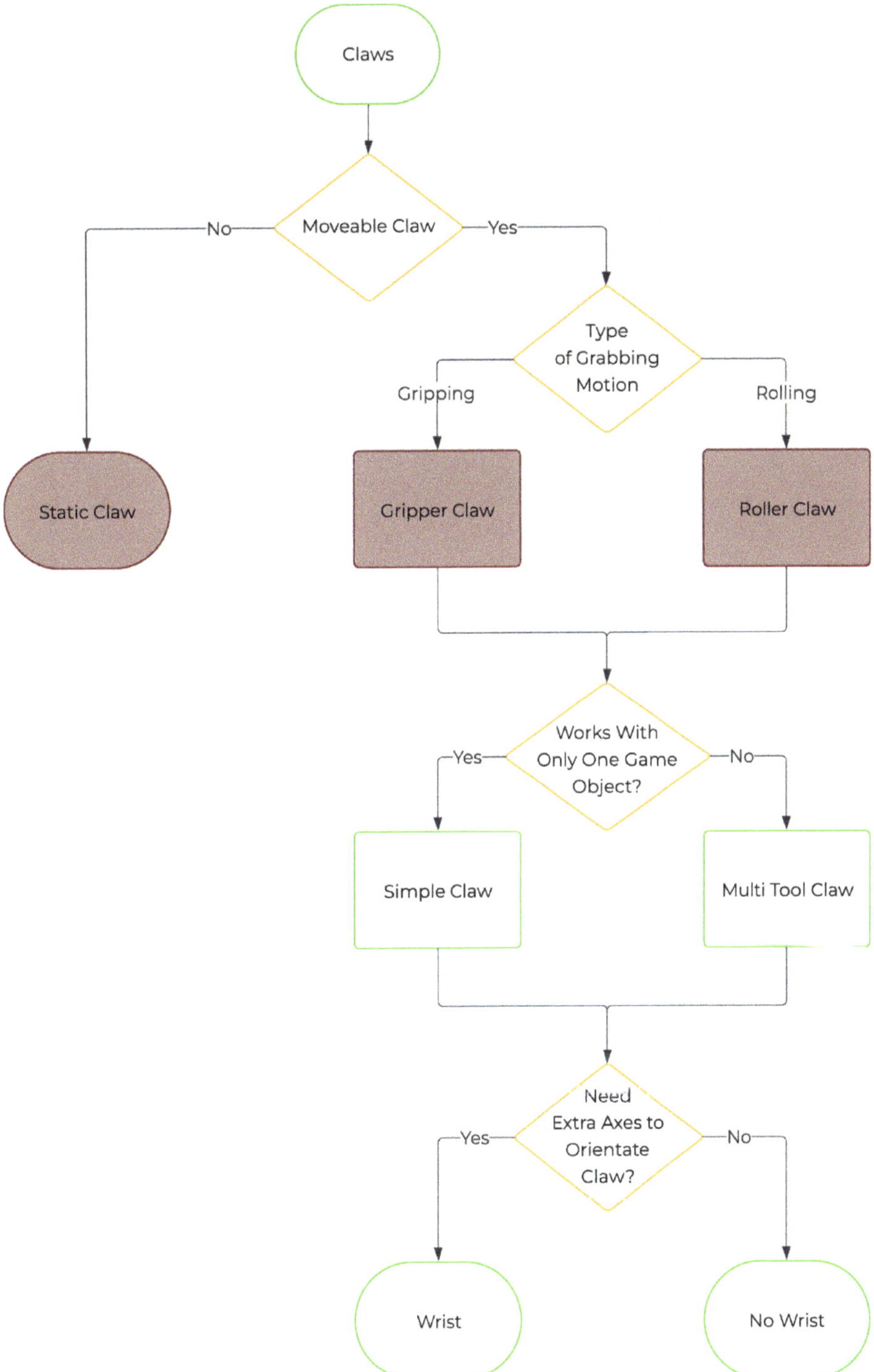


Diag. 10.21: Claws Flowchart

# 11. Advanced: Motor Sharing Techniques

## 11.1 Introduction

Motor sharing techniques use structural mechanisms that can be integrated into the design of all the mechanisms discussed in this handbook thus far. They are particularly useful when competition rules limit the number of motors that are available. While the modern power distribution panels and hubs have enough motor ports rendering a ratchet or a differential structure unnecessary, it is still good to know how such structures can be used to get more functionality out of fewer motors.

Motor sharing techniques complicate mechanisms, so they should only be used when truly necessary on a competition robot design. However, for experiential and prototyping endeavors during the offseason, it may be worthwhile to try using such designs.

## 11.2 Ratchets

In its most basic form, a ratchet allows the output of a gearbox or motor to spin in one direction, while preventing motion in the other by locking the mechanism.

Diag. 11.1: Ratchet Ex. 1 Side View; Ref. 39

In this ratchet design, there are 2 key pivots shown by the black dots. In each set the ratchet gear is on the left, while the ratchet pawl is on the right. When the gear rotates counter clockwise, the pawl lifts up and clicks over each tooth, as shown in the top right of diagram 11.1. When the gear tries to rotate clockwise, the pawl catches on the gear's tooth and the gear stops turning.

To keep the pawl in contact with the gear, it must be tensioned, typically by a spring, to function properly. Rubber cords or rubber bands can also serve as tensioners, although they are less durable and reliable compared to springs. Getting the level of tension right is a balancing act. Inherently, some energy is lost to friction due to the resistance between the gear and the pawl, reducing the efficiency of gearboxes that incorporate a ratchet. If the tension on the pawl is too strong, this frictional loss is increased unnecessarily. If the tension is too weak, there will be a delay in the clamp down on the gear to stop movement, or even a failure to do so. How much to tension the pawl depends on the application of the ratchet.

A single ratchet can be used in a few ways. All modern bikes have ratchets so that the biker does not need to continue pedalling and the back wheel is free to spin forward on its own. A similar application of this concept can be found in a flywheel mechanism.

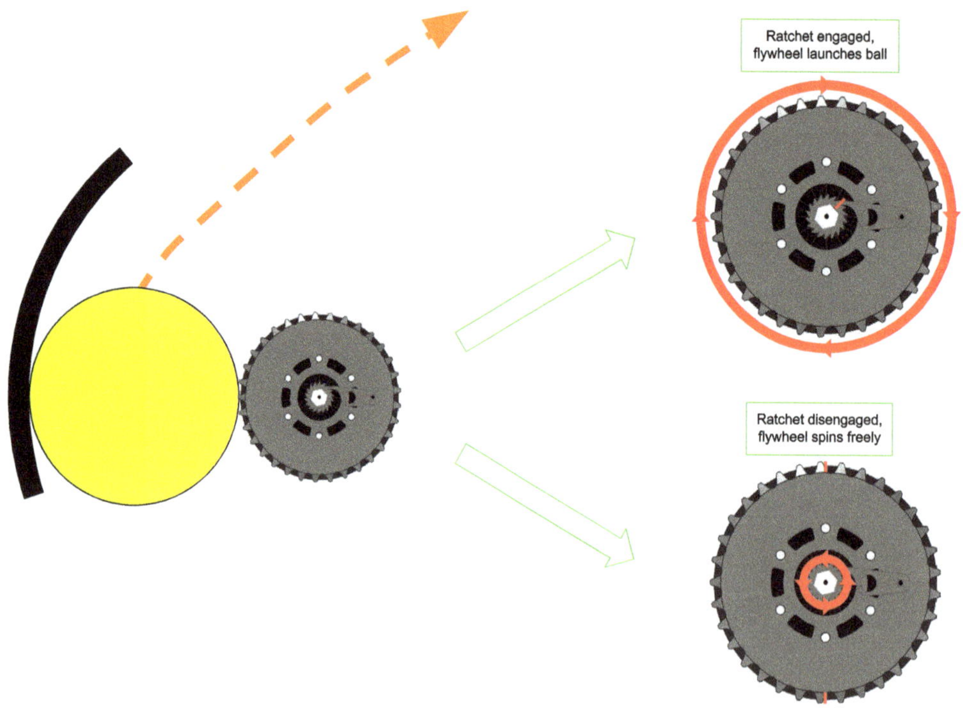


Diag. 11.2: Ratchet Flywheel Abstract Ex.

In this abstract diagram of a single vertical flywheel, a ratchet is added as part of the design. Assume that the ratchet mechanism structure is mounted onto the plate sprocket. The gear of the ratchet structure and the flywheel both turn with the axle. However, the plate sprocket turns independently of the axle since it has a bearing hole bore. Looking at the top right of diagram 11.2, when the sprocket is driven clockwise, the ratchet locks, engaging the ratchet gear. This turns the axle clockwise which then turns the flywheel clockwise launching the ball. In the bottom right of diagram 11.2, when the sprocket is no longer being driven, (or even if it spins counterclockwise), the ratchet pawl simply clicks over the ratchet gear, which allows the flywheel to continue spinning clockwise.

The flywheel thus keeps its momentum even when it is not being powered by its motors. A potentially useful application of this would be if the game object is exceptionally heavy. In order to design the flywheel to transfer enough energy to fire a heavy game object, energy could be stored in a heavy plate. However, keeping motors running constantly to power a heavy flywheel throughout a match would drain the robot's battery. So instead, the flywheel can be periodically powered by the gearbox while the ratchet allows the flywheel to continue to spin freely between power inputs, thereby maintaining momentum.

It's important to note that the ratchet causes friction, so the flywheel will eventually slow down after being spun up. Nonetheless, the deceleration takes significantly longer than the instant stop that occurs when the gearbox is disengaged.

Notice how in the abstract example, the flywheel is only being powered when the motor spins one direction. A thoughtful engineer might then ask: what could the motor do when spun the other direction? This is where ratchet structures can be used creatively to enable a single motor to perform different actions depending on the direction of its rotation.

Another way to use ratchets would be to power 2 different arms with the same motor.

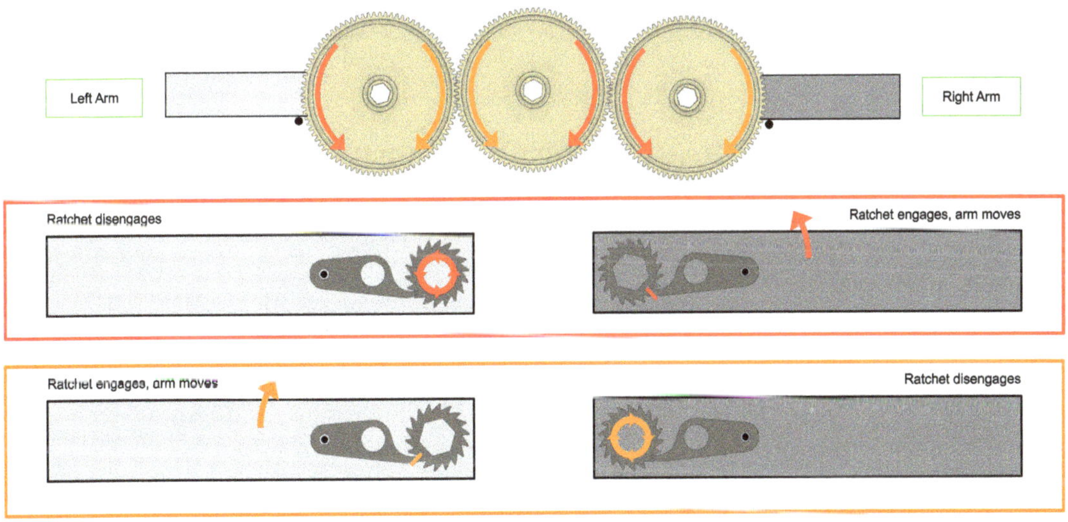


Diag. 11.3: Ratchet 2 Arms Abstract Ex.

This abstract diagram shows 2 different arms with end stops, represented by the small black dots just under the arms. The red box shows how the ratchets, hidden underneath the large gears, interact with the arms when the gears spin in the direction of the red arrows. The orange box shows the same interaction for the opposite direction, indicated by the orange arrows. Rotating the gears following the red arrows, the left arm's ratchet simply clicks over without moving the arm. Meanwhile, the right arm's ratchet engages and powers the right arm. The reverse happens when the gears spin in the direction of the orange arrows.

When the ratchets are disengaged—meaning the gears spin in the opposite direction of the "engaging" direction—the arms return to their resting position under the force of gravity. In practice, gravity can be replaced with a passive spring or rubber band to retract the arms in a specific direction. This setup allows the arm to be fully powered when moving in the ratchet's engaging direction, and passively pulled by a spring or rubber band to the resting position when the ratchet is disengaged.

In doing so, a single motor spinning this set of gears can effectively power 2 different arms. The movement of the arms are synchronized in the sense that one arm would have to start to spin back down to the end stops while the other arm is rotated away from the end stops. The difference between this setup and simply bolting the arms directly to the gears is that, with the ratchet system, one arm can continue rotating independently while the other arm only starts to move when the rotation direction is reversed. Potentially, this type of mechanism could be used in a filter to open and close gateways to direct game objects through different pathways within the robot.

This ratchet design could be modified to replace one of the arms with a flywheel ratchet. How this would work is that the flywheel would be powered when the motor is spun in one direction, while the arm would be powered when the motor is spun in the opposite direction. This extends the functionality of the flywheel ratchet example in diagram 11.2 as the motor can now drive a separate arm mechanism instead of just remaining idle when not powering the flywheel.

A real-world example of a ratchet design in an intake is taken from Reference 1. See https://www.frc-handbook.com/ref/1 for more information on the game and full robot.

Diag. 11.4: Ratchet Intake Ex. Isometric View; Ref. 1

At the time the team had exhausted its budget for additional motors and pneumatics. This posed a design challenge: to develop an intake that can start within sizing restrictions, flip down to become an overbumper intake, and operate using a single motor.

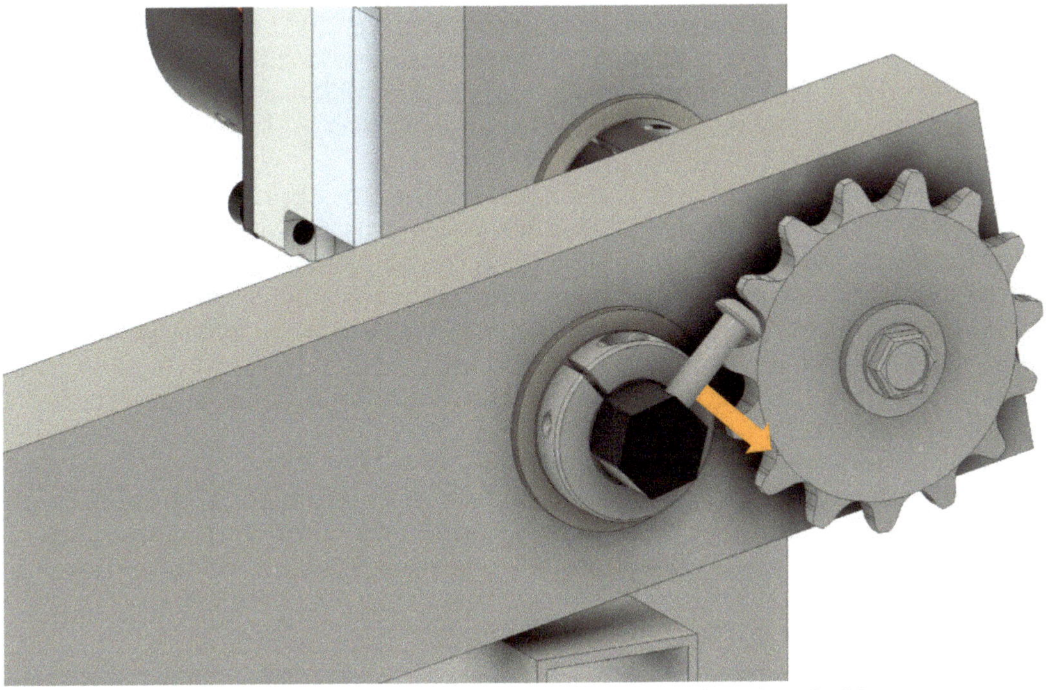

Diag. 11.5: Ratchet Intake Ex. Zoomed In View; Ref. 1

This ratchet design works by using the head of a screw as the ratchet's pawl. The screw is threaded into a tapped hole on the side of a Hex axle. The sprocket acts as the ratchet gear in this structure. The screw was tensioned to rotate clockwise passively with a rubber band, shown by the orange arrow. This means that the pawl simply clicks over the sprocket when spun clockwise, and the ratchet is engaged when spun counterclockwise.

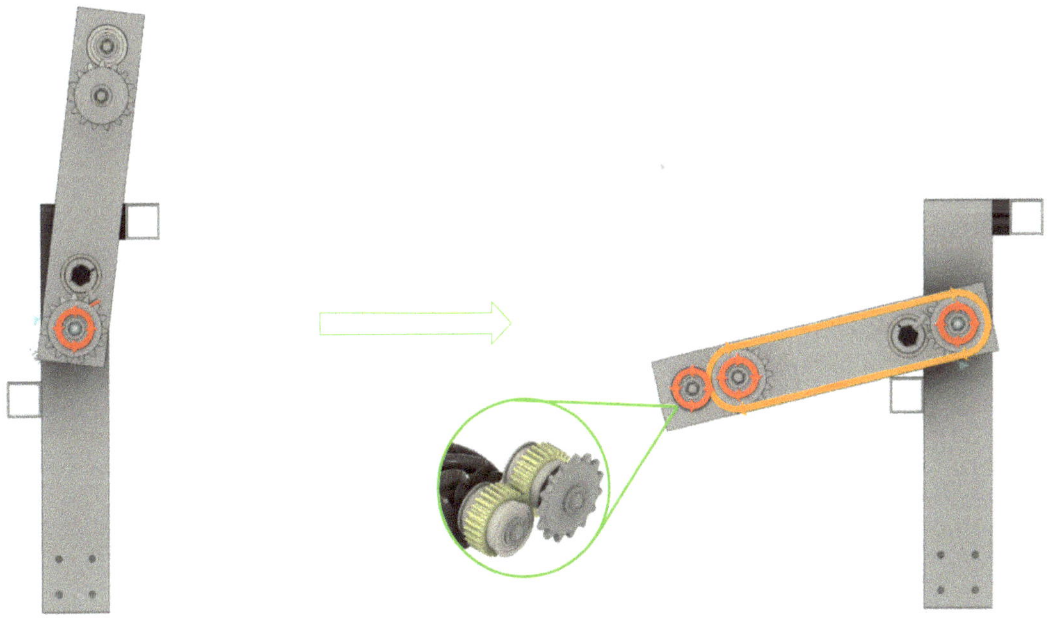

Diag. 11.6: Ratchet Intake Ex. Side View; Ref. 1

The intake starts in the left position of diagram 11.6 at the start of an FRC game. When the motor spins the driving sprocket counterclockwise, the ratchet is engaged, pushing the arm to rotate counter clockwise. Now the arm is flipped down per the right position of diagram 11.6. When the motor spins clockwise, the sprocket clicks over the pawl, allowing a chain to turn to drive the other sprocket. As the zoom in view indicates, gears inside the stock reverse the direction of the power transmission so that the final driven axle spins counter clockwise. From this view, spinning the driving axle clockwise powers the intake's axle in the correct direction to pull balls into the robot.

A limitation of this design is that the intake cannot eject balls as the ratchet prevents the intake from running in reverse. In theory, this should not be necessary since a good intake design should be able to pull game pieces into the robot without jamming and the driver can choose when to intake pieces. In practice, this intake had a few design flaws. The custom ratchet, made using a bolt, was a cheap solution that wore down over time and required regular maintenance to remain functional. If budget had allowed, incorporating a commercial ratchet structure would probably have been a better choice.

While only a few examples have been shown in this chapter, the applications for ratchets are limitless. Ratchets provide an alternative approach to designing mechanisms that operate with fewer motors. In general, it is better to use more motors if the motors are available, as ratchets complicate mechanisms and cause friction. However, there are times when using ratchets make sense.

## 11.3 Differentials

Differentials are a distinct type of motor-sharing mechanism. The basic principle involves using 2 motors to jointly power 2 separate outputs. When both motors spin in the same direction at the same speed, one output is powered. When the motors spin in the opposite directions at the same speed, a different output is powered. If the motors spin at different speeds, both outputs can be powered at the same time, though neither mechanism would be fully powered.

Practical examples of differential mechanisms are few in FRC. Differentials were more commonly used in VRC when the number of motors allowed was rather restrictive. Even so, differentials are used only by the most experienced VRC teams, as differentials are complex mechanisms.

### Differentials to Power an Arm on A Lift

Here is a theoretical application of differentials for FRC. This differential allows for both an arm on a lift to be powered by 2 gearboxes at the bottom of the lift.

In diagram 11.7, assume that the red line represents a pulley wound all the way through the lift mechanism. The movement of the lift mechanism is one of the outputs of this differential. The orange circle represents the other output, a pulley, that then drives a pivot, like an arm. The green and cyan circles represent 2 different pulleys that are the 2 inputs for this differential system. The remaining yellow and black dots represent additional pulleys that create the pathway of the belt and are all free spinning. Now to describe what is happening in each of the different frames F1 to F4:

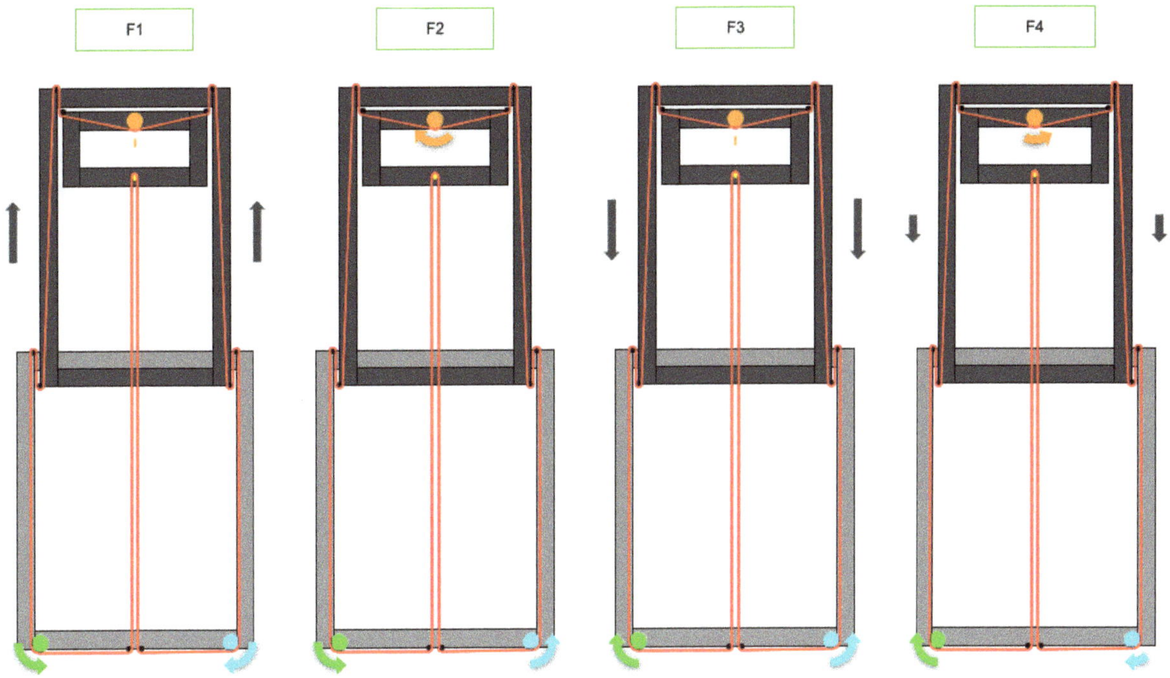


Diag. 11.7: Differential Abstract Ex. Front View

1. In F1, when the green pulley and the cyan pulley spin in opposite directions, with the green pulley spinning counter clockwise, and the cyan pulley spinning clockwise, the belt is pulled to tension, locking the orange pulley. The weight of the lift now rests on the non-rotating orange pulley as the 2 lower driving pulleys move the lift up.

2. In F2, the green pulley and cyan pulley spin in the same direction, counter clockwise. This drives the belt through the entire system which spins the orange pulley clockwise.

3. In F3, when the green pulley and the cyan pulley spin in opposite directions, with the green pulley spinning clockwise, and the cyan pulley spinning counter clockwise, the belt is pulled to tension, locking the orange pulley. Tension is increased around the yellow pulley while tension is released around the orange pulley, resulting in the 2 lower driving pulleys moving the lift down.

4. In F4, a difference in the inputs allows both outputs to activate at the same time. When the green pulley and cyan pulley are spinning in the same direction, clockwise, but with the cyan pulley spinning at a slower speed, both the orange pulley spins and the lift lowers at the same time. However, the orange pulley's rotation and the lift are slower and their load capacity is reduced, compared to when only one output receives all of the power from both inputs.

Variations in different split input modes can be calibrated for specific differential designs.

### Differential Swerve Module

Another theoretical application is for a differential swerve module. (See Subchapter 2.8 on how a swerve drivebase works.) In theory, the combined power of both motors in a swerve module can be directed solely into driving the wheel, as long as the module itself is not rotating and the wheel is aligned in a fixed direction.

A few differential swerve module designs have been tested by experienced teams only to find that there is little practical performance improvement.

The problem likely stems from the fact that a higher performance drivebase is expected to move faster. If the swerve is geared to move optimally at a high speed when both motors are powering the drive wheel, the swerve module will not function when the inputs are split between powering the drive wheel and rotating the wheel's heading. This then leads to the logical decision to gear differential swerve modules to drive optimally when the inputs are split. However, once this is done, the differential swerve is not any faster than a regular swerve module.

One advantage of introducing differentials into a swerve module is that it increases the maximum torque of the drivebase. Such a design could be useful in a pushing contest, but a number of things like the friction between the wheel and the ground and the weight of the robot need to be considered to make use of the additional torque.

Do note that differential swerve drive modules discussion is now moot following the introduction of the 2024 FRC rule limiting propulsion to four motors in the drivebase. Since a standard differential swerve system requires two propulsion motors per module, a full four-module drivetrain would necessitate eight motors—rendering it non-compliant with current regulations. This discussion is included here to promote a deeper understanding of differential mechanisms.

### Creative Use of Differentials

There are many more ways to use differentials to power 2 different outputs with 2 different inputs. Researching online and perhaps taking inspiration from VRC designs will lead to some interesting discoveries.

## 11.4 Gear Shifting

Gear shifting with pneumatics is another way to increase the functionality of a robot when working with a limited number of motors.

Gear shifting is most commonly applied within the gearbox of a West Coast Drivebase, (see Subchapter 2.9 on Drivebase Gearbox Shifting). This allows teams to extract greater utility from fewer motors, enabling the drivebase to run at 2 different speeds: a high torque pushing mode and a high speed maneuverability mode.

However, gear shifting can also be extended beyond the drivebase to connect multiple mechanisms in a robot. By selectively engaging different outputs, a single gearbox can alternate between powering distinct subsystems—further optimizing motor usage.

Refer to https://www.frc-handbook.com/ref/20 for an example from Team 254.

Diag. 11.8: Shifting Drivebase/Climb Ex. Isometric View; Ref. 20

The full complexity of this gearbox is not immediately evident from the single image provided in Team 254's technical binder. This setup is not just a 2-speed gearbox that uses dog gears to shift between high speed and high torque for the drivebase. On top of that, there is another shifter, that switches the gearbox between powering the drivebase and powering the climbing mechanism.

Using gear shifting to power 2 different output axles is a clever way to get more value out of fewer motors in a design. This example makes use of gear shifting to power 2 distinct mechanisms.

## 11.5 Summary on Motor Sharing Techniques

The more advanced motor sharing techniques shown in this chapter allow teams to create some really dynamic mechanisms that operate with fewer motors. That said, it is always better to use a simpler mechanism to accomplish a task whenever possible. Only use the motor sharing techniques described in this chapter when there are no other options.

| Ratchets |
| --- |
| • Pawl of a ratchet allows rotation of an axis in one direction and locks the mechanism in the opposite direction<br>• Typical applications involve spinning the motor in one direction to power one mechanism and spinning in the opposite direction to power another<br>  ○ Applies to a double ratchet system<br>  ○ Can use the lock of the ratchet as a mechanism action (like flipping down an intake)<br>• A separate application allows a mechanism to operate without the the need for constant motor power, typically used with a heavy flywheel |
| **Differentials** |
| • Two motors jointly power 2 separate outputs<br>• Power is shared between the outputs, either at full power to a single mechanism, or splits total available power between both mechanisms |
| **Gear Shifting** |
| • Pneumatics allow for the engaging of different dog shifting gears<br>• Allows the same gearbox to switch effective outputs to power different mechanisms |

Diag. 11.9: Motor Sharing Summary Table

Motor sharing techniques are extremely situational and should only be used when there are limited motors available. The following flowchart is based on the assumption that the team needs to use a motor sharing technique to either compensate for limited funds or limited motor ports.

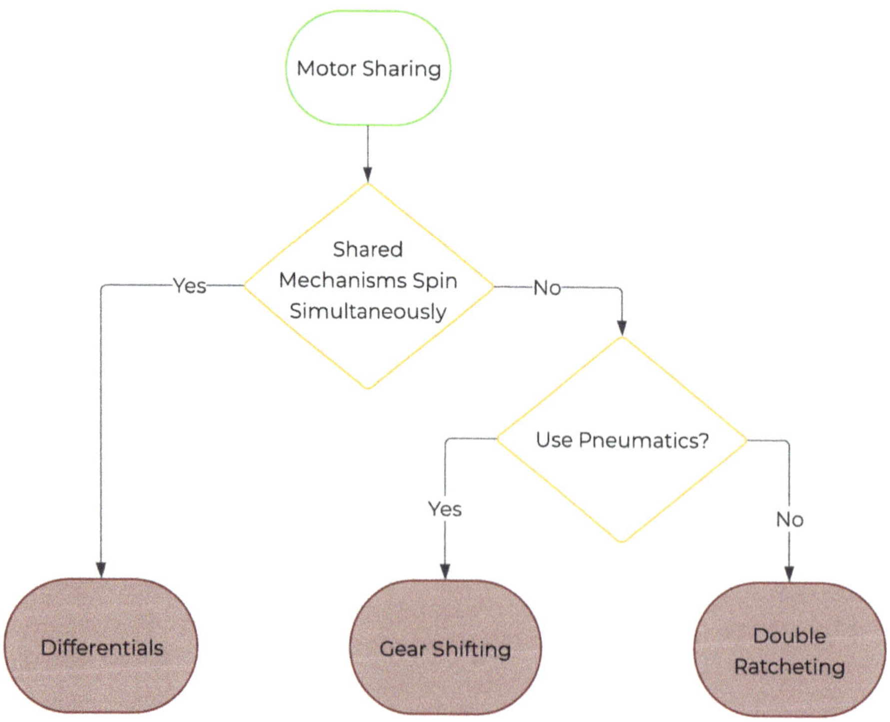


Diag. 11.10: Motor Sharing Flowchart

# 12. Design Process

## 12.1 Introduction

This final chapter discusses the overall design process and the focus will predominantly be on the Kickoff Design Meeting that is attended by all members.

## 12.2 Navigating the Kickoff Design Discussion

With a solid understanding of general mechanisms and design principles, the most pivotal moment in the early FRC season is the Kickoff design meeting. This is where imagination, strategic insight, and technical knowledge come together to chart the course for the build season.

### Aim for a Complete Design at Kickoff

While some teams consider the Kickoff design discussion to be more of an exploratory exercise, it is best to use this session to create as complete a design idea as possible from the get-go. Given that the build window is 2-3 months, time is of the essence! A solid, well-thought-out plan from day one can provide the clarity and direction needed to keep the team on schedule.

Nonetheless, new design ideas may continue to surface later on in the season. Robot in 3 Days (RI3Ds) are published in the first week of the FRC season by non-competing University teams. They may inspire a new robot design. It may even be tempting to wait for RI3Ds for the initial inspiration. However, it is best to apply creative brainstorming before any preconceived solutions are considered as all creativity could be starved off after viewing somebody else's design which may not consider the team's specific strategy (which will be discussed in this chapter).

Similarly, a new superior idea may pop up during early CAD work and tempt the team to pivot.

While flexibility to incorporate new ideas during the build season can be beneficial, it's essential to weigh the costs carefully.

### The Risks of Redesign

Altering your design mid-season can range from a small tweak to a complete overhaul. A full pivot could mean restarting the entire CAD process; a partial redesign may involve remodeling just a single mechanism. In either case, the impact on the CAD sub-team's workload and the broader production timeline is significant. Even when changing a single mechanism, extensive rework in integration is necessary. Delays at this stage cascade into building, wiring, programming, and drive testing, potentially jeopardizing your performance at your first regional or district event.

### Encouraging Early Creativity

A way to circumvent a complete or partial redesign is to get all of the creative juices flowing in that first Kickoff meeting. Bolstered by everyone's excitement and fresh perspective on the game challenge, that's when creativity should be at its peak.

If a solid design is chosen from the get go, the team can be comfortable sticking with the original design plan, even if slightly better ideas appear on the internet later on.

**When to Pivot**

That being said, there are times to restart the CAD design. For example, if new ideas or strategies provide substantial competitive advantage against the drawback of a delayed build timeline, it may be worthwhile changing course.

Sometimes despite a team's best efforts, a design idea on the drawing block proves to be impractical late in the season. A late-stage failure leaves the team with no choice but to re-design what is simple enough to get done in time instead of what is most optimal for playing an FRC game.

**Mitigating Late-Stage Failures**

Knowing what works and what doesn't work is the obvious way to circumvent this problem. Building a deep understanding of common FRC mechanisms as detailed in this book is a good start.

A less obvious piece of advice is to never hesitate to admit uncertainty. If the team is unsure whether a mechanism will work, or whether the team has the capability to build a specific mechanism, prototyping is a safe way to test ideas. Design is very theoretical, while what matters most is what actually works. Prototyping turns theoretical designs into practical insights. It's a low cost, high-reward way to evaluate risk before committing to full-scale production. (See the end of Subchapter 12.5 for more details on prototyping.)

**Designing with Strategy in Mind**

This chapter outlines a practical and strategic approach to Kickoff design planning—one that aims to preempt the common challenges of mid-season redesign and late-game disappointment. Remember, designing an FRC robot is not just about technical know-how. It's about applying strategic thinking, fostering team collaboration, and remaining flexible without being reactionary.

As you enter your next Kickoff meeting, come equipped not just with ideas, but with insight. Be bold in your imagination, yet grounded in your methods.

## 12.3 Understanding the Game and Its Context

Every FRC season begins with the game reveal and the release of the official Game Manual. Before any design work begins, it is important that every team member study the Game Manual assiduously and thoroughly understand the game—not just the rules, but the structure and strategic nuances that will influence robot function.

**Match Structure and Alliance Dynamics**

FRC matches are contested between two alliances, each composed of three different teams. This format requires each robot to not only compete effectively but also to collaborate and synergize with other alliance partners.

FRC events begin with a series of qualification matches, followed by a double-elimination playoff bracket.

Team rankings during qualification rounds are determined primarily by average Ranking Points (RPs), which are awarded for match outcomes —3 points for a win, 1 for a tie—and for completing specific bonus tasks.

During the elimination playoffs, the ultimate standing of the team is no longer determined by RP but by match points.

A key nuance is that the tasks granting RPs are not always the same as those maximizing match points. For example, the games in some years may reward high-point scoring for win conditions, while bonus RPs may depend on smaller, strategic side objectives. Recognizing this distinction can help your team prioritize effectively.

### Judged Awards and Pit Interviews

Beyond gameplay, there are additional FRC awards that recognize excellence in areas such as design, innovation, and team dynamics. Winning these awards requires not just a well-performing robot but a team that understands and can articulate its design rationale.

Judges may speak to any member during the Pit Interview. It is essential that every team member is able to explain the robot's features and defend the team's design choices. If you've read this handbook closely, you're already contributing to this goal.

## 12.4 Formulating a Game Strategy

With the rules and structure of the game clearly understood, the team can move on to the strategic layer of robot design.

### Setting a Goal

Different teams have different strengths. Strategy must reflect both game mechanics and team capabilities. There are a few goals a team can build a strategy around.

1. **Maximize Ranking Points.** The first eight teams with the highest ranking points (RP) from the qualifying rounds gain the privilege of selecting partners during alliance selection. (For how the alliance selection process works see https://www.frc-handbook.com/volI/12/selectionprocess.) As higher ranking teams are likely to choose amongst the top eight, the objective is to vie for the top twelve spots to have an opportunity for this privilege. Coupled with strong scouting capability to select complementing teams in terms of scoring potential or ability for the playoff matches, alliance captains thus can improve the odds of winning at the elimination playoff. A ranking point focused robot is one that prioritises completing all side missions that result in bonus RPs.

2. **Maximize Match Points**. Since what matters in the final elimination bracket is simply winning matches by outscoring the other alliance in match points, focusing on match points can be a sound strategy. This strategy may very well result in a team placing at the top of the qualification rounds since the winning teams in the qualification matches are also awarded RPs (see Subchapter 12.3). Even if the team doesn't rank very high, the match point focused robot will likely catch the eye of the top teams' scouts and be invited to form an alliance for the playoffs.

3. **Enable Alliance Partners.** Whether this is a viable strategy is dependent on the FRC game. Some years there are specific scoring or "activating" niches that partner teams can fill. An example would be the 2024 FRC game where if a partner robot places 2 notes in the amp, the human player "can hit a button to amplify the Speaker," which results in notes scored in the Speaker being worth 5 points instead of 2 points for the next 10 seconds. Most top tier teams that year built robots to efficiently cycle Speaker shots, creating a niche for a partner robot to efficiently cycle amp shots to keep the Speaker amplified for most of the match. Such a strategy enables partner teams, though the specific team employing this strategy would not be able to score many points individually. If this goal is viable for a specific FRC game, "weaker" FRC teams can harness it for a chance to make it to the playoffs. While it is unlikely that a team employing this strategy will be a first pick in the alliance selection process, a team that stands out as being able to enable other teams has a decent chance of being a second pick.

4. **Play Defense.** For most FRC games, hampering opposing robots in some way is a viable strategy to limit the scoring of the opposing alliance. Often there are limitations in the rules on how a robot can play defense against the opposing alliance. Strategies focused on this goal involve identifying the strategies other teams and alliances are likely to employ and figuring out how to disrupt said strategies without breaking the game rules. Like goal 3, this strategy is a good goal for a "weaker" FRC team to build a game strategy around. Again, it is unlikely for a defense focused robot to be chosen in the first round of the alliance selection process, but it is a niche that makes a team stand out for the second round of selections.

5. **Balanced or Hybrid Approach**. The 4 goals above are not mutually exclusive. That being said, most robots are primarily designed around one goal, with the rest being secondary goals.

### Translating a Game Strategy into Design

Regardless of the goal in mind, the next step is to plot the robot's ideal match flow. Map its movements: how it begins, cycles through scoring, and finishes the match. Be specific about actions and locations. Make notes on when and how game objects are picked up, the scoring roadmap and the robot orientation.

To provide examples of what these types of strategy diagrams look like, here is a case study taken from Team 4817's strategy for the 2024 FRC game Crescendo.

In this game, two alliances, each comprising three teams, compete to score points by manipulating game pieces called "Notes"—orange foam rings—into various field elements:

- **Speaker**: A high goal where robots could shoot Notes for higher point values.
- **Amp**: A lower goal allowing robots to deposit Notes.
- **Trap**: An endgame element where robots could place Notes.

There is also a central structure on the field known as the Stage, equipped with chains for robots to climb during the endgame phase.

For the game reveal video and the full game rules refer to https://www.frc-handbook.com/past-frc-games.

For context, 4817 employed a balanced strategy, where the primary goal was to maximize ranking points (RPs), with a secondary goal to maximise match points and and a backup goal to play defense.

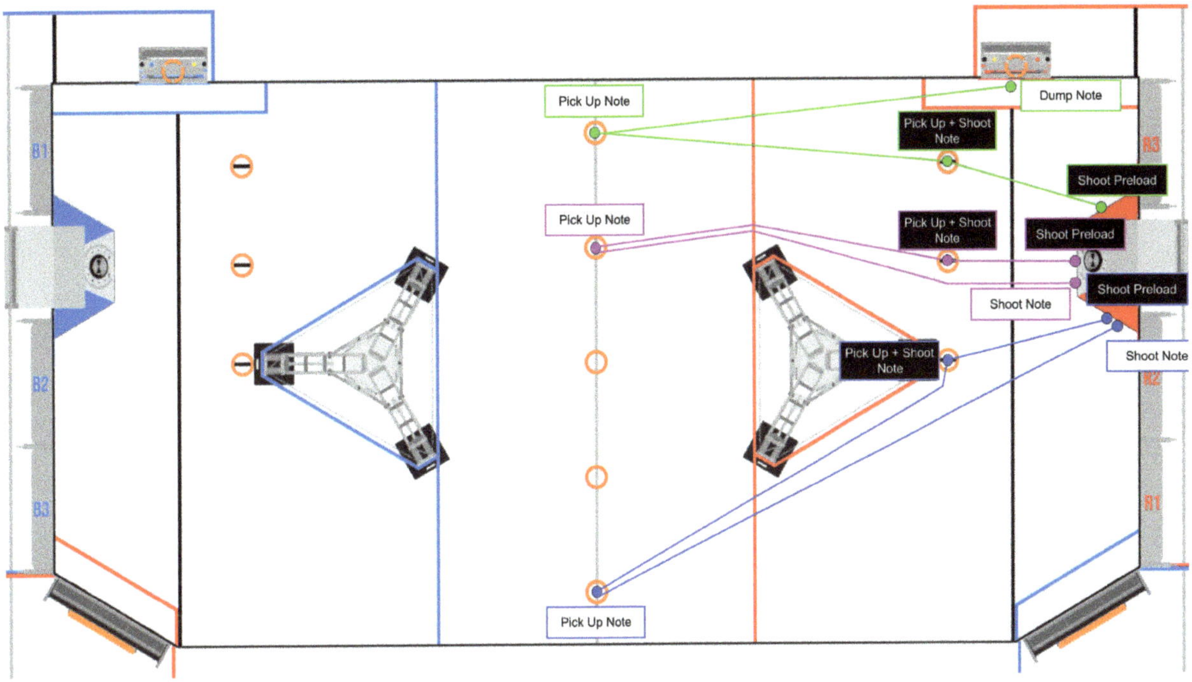


Diag. 12.1: Match Start Strategy, 2024 FRC Game

This match strategy diagram shows 3 different pathways that Team 4817 could start a match. There is an autonomous section labelled in black boxes, and a driver control section labelled in white boxes. This diagram represents the team's strategy prior to finalising the robot design for that year. It turned out later that more could be achieved during the 15 second autonomous period. Do note that the initial strategy need not be the final game strategy.

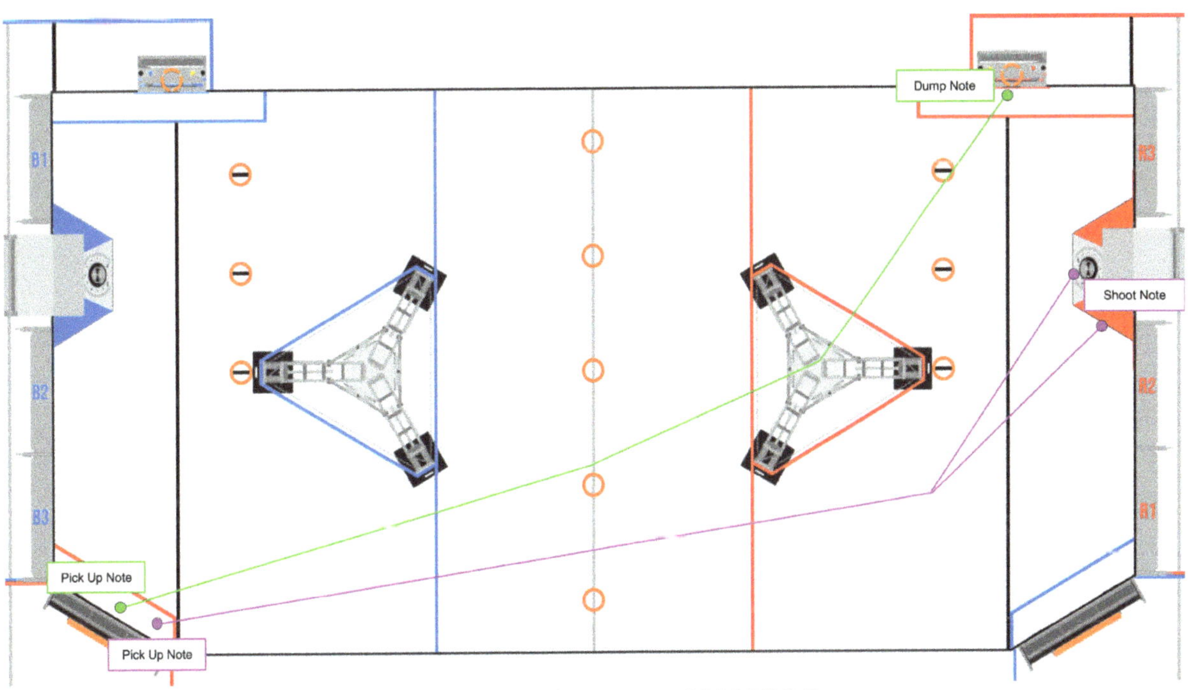


Diag. 12.2: Match Cycle Strategy, 2024 FRC Game

In diagram 12.2, the green path shows the cycle route for dumping notes into the Amp, while the pink paths show the 2 cycle options for shooting notes into the Speaker. The more cycles Team 4817 could complete independently, the more likely that Team 4817 can aid an alliance in achieving the Melody Bonus RP. Therefore, a robot that moves fast, feeds notes efficiently into the Amp and shoots notes accurately from multiple angles into the Speaker is ideal.

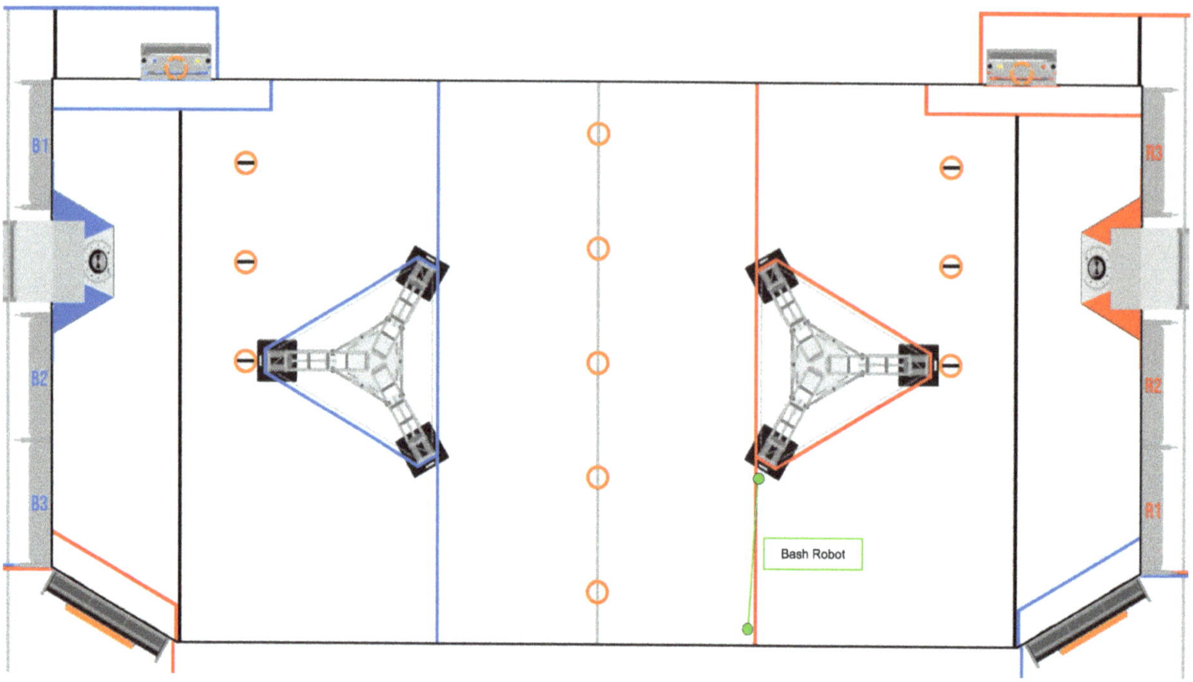


Diag. 12.3: Match Defense Strategy, 2024 FRC Game

While Team 4817 did not plan to primarily play defense for the 2024 game, in the event the robot's non-drivebase mechanisms were damaged, this would be an alternate strategy to employ.

The green path shows the best choke point to patrol without attracting penalty points which are given for contacting opposing robots in penalty zones as marked out by the center blue triangle and the blue lines. This defense route is partly determined by the best route to hamper a mirror of the cycle routes shown in diagram 12.2. Even though such a defense isn't foolproof as opposing robots could go around such a path, there is the benefit of forcing the opponents to take a longer route, and if needed the defense route could be adapted in real time to impede other strategies.

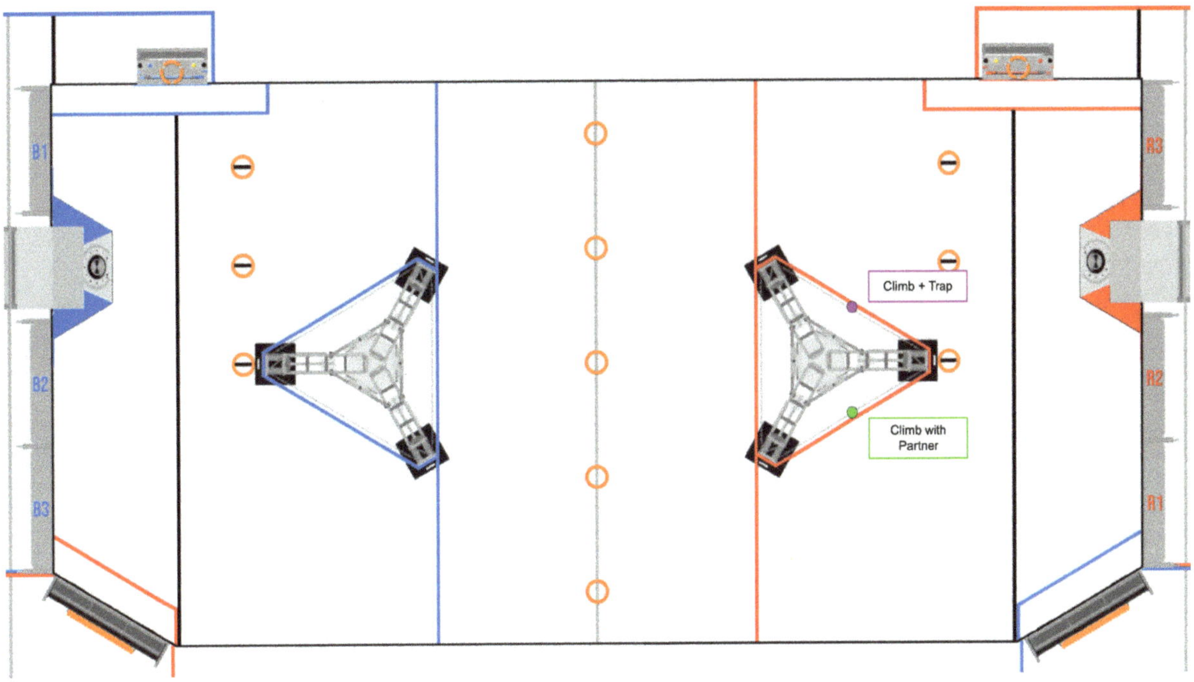


Diag. 12.4: Match End Strategy, 2024 FRC Game

In order to improve the chances of getting the Ensemble Bonus RP in each qualifying match, Team 4817 wanted to maximize the stage points achieved at the end of games. That translated into a design capable of both scoring in the trap and climbing with a partner robot, providing the flexibility to support the alliance in scoring sufficient stage points.

Reviewing all 4 strategy diagrams together, it was determined that the required capabilities of the robot were:

1. To have a fast drivebase
2. To be durable enough to play defense
3. To be able to shoot into the Speaker from the close starting locations of the notes
4. To be able to score in the Amp
5. To be able to climb with a partner
6. To be able to climb and score the Trap

This strategy led to the robot design for Reference 8. More details can be found at https://www.frc-handbook.com/ref/8.

The purpose of these strategies are to define the required capabilities—and those needs, in turn, shape the robot's design. A game strategy diagram doesn't necessarily explain how the robot scores RP or match points in various ways, but it helps focus the team on what engineering challenges need to be solved.

Breaking into smaller groups can be an effective way to generate a variety of strategies, but the team must converge on a single unified approach before proceeding further.

This imagined cycle should be grounded in the team's capabilities and budgetary constraints. For example, if a complex endgame climb yields only marginal point gains, it may not be a worthwhile pursuit for a less experienced team. A team that adopts the "maximize match points" strategy need not attempt the complicated climb that awards only 4 additional match points if team members do not have the technical knowhow. It is far better to focus on building a reliable robot that can consistently score with game objects than building an unreliable robot with lots of bells and whistles. An FRC team should not go beyond its limitations and be realistic in what its robot is able to do in this "idealized" strategy.

## 12.5 Planning the Robot

In this phase, all of the information in this handbook is put to use. The team now knows what the robot needs to be able to do. It is time to start sketching out robot designs that incorporate the various mechanisms described in this handbook.

Using collaborative tools such as Miro or Jamboard, develop three 2D views (top, side, front) sketches that roughly show the robot's geometry, mechanism layout, and basic dimensions. The more effort put in here during planning, the less time will be spent dealing with integration issues later.

Here is an example of what one set of planning sketches look like.

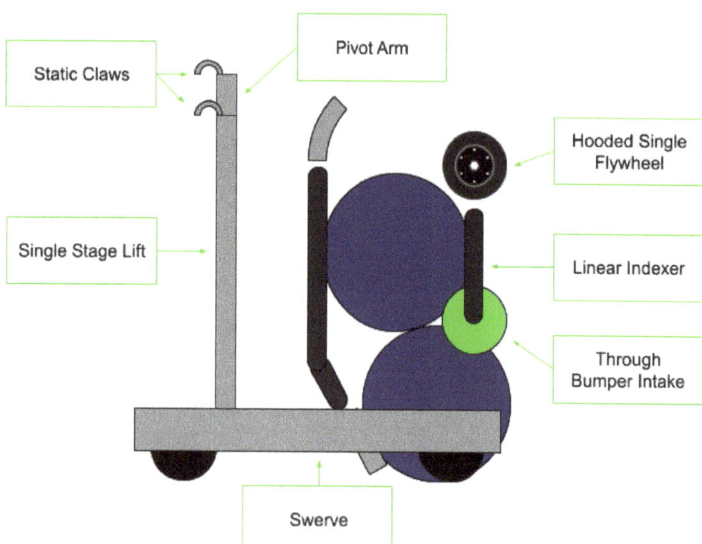


Diag. 12.5: Master Planning Sketch Side, 2022 FRC Game, Ref. 12

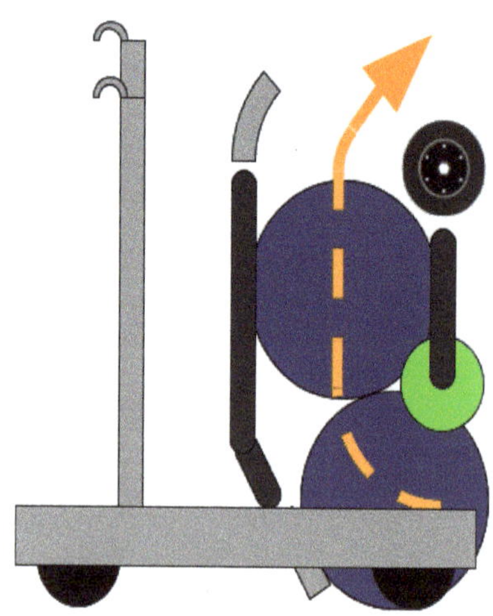

Diag. 12.6: Master Planning Sketch Ball Path, 2022 FRC Game, Ref. 12

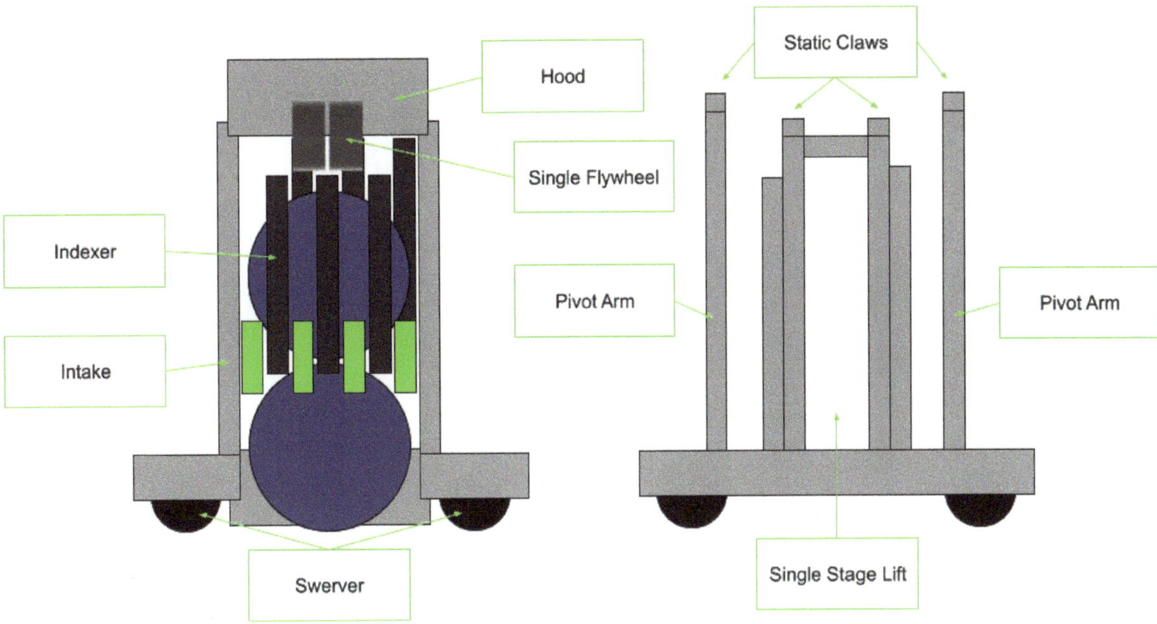


Diag. 12.7: Master Planning Sketch Front and Back, 2022 FRC Game, Ref. 12

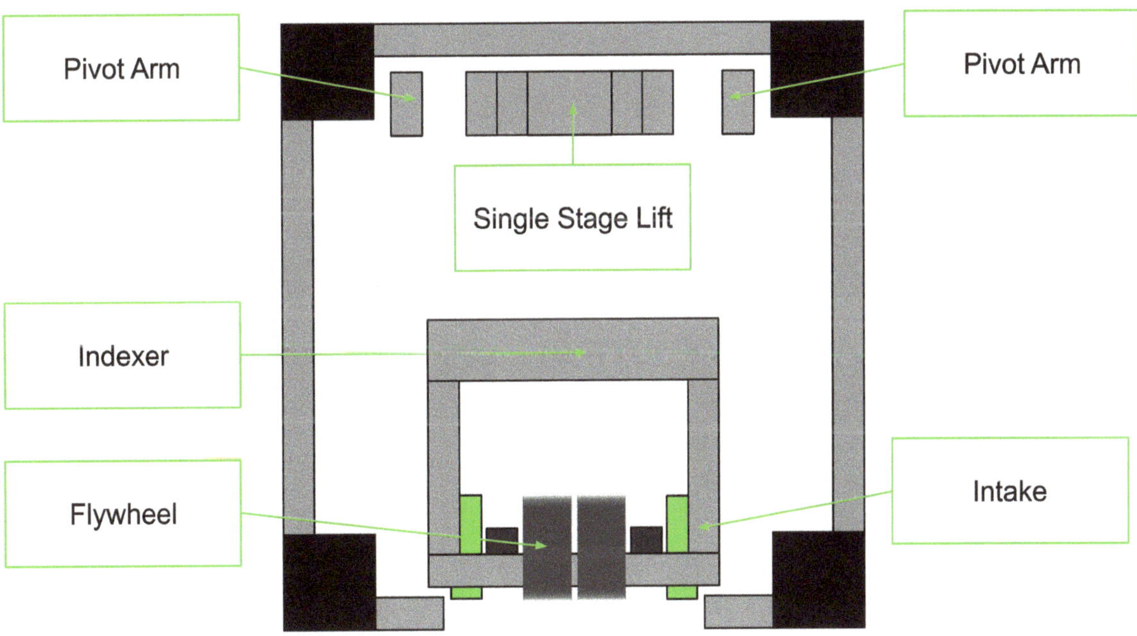


Diag. 12.8: Master Planning Sketch Top, 2022 FRC Game, Ref. 12

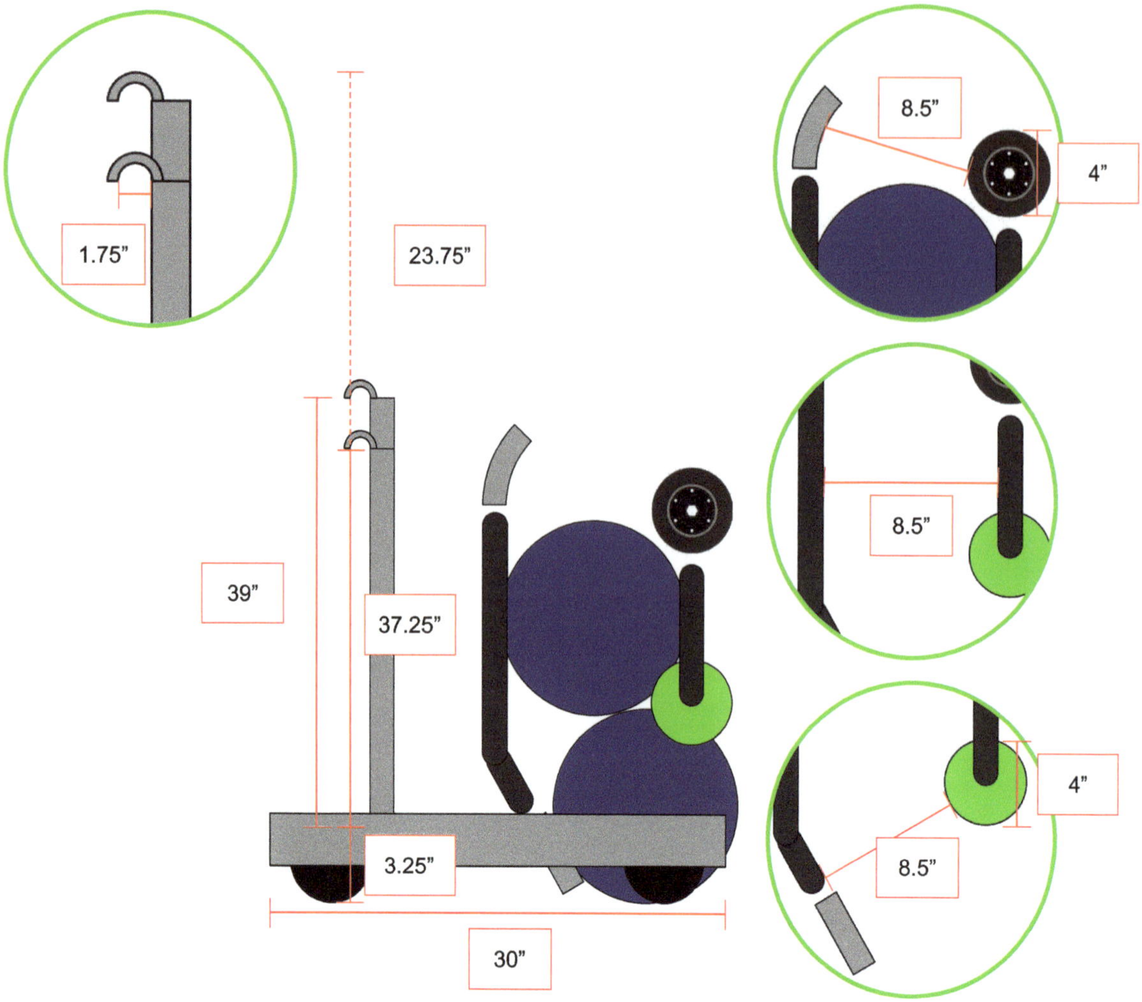


Diag. 12.9: Master Planning Sketch Side Dimensions, 2022 FRC Game, Ref. 12

As can be seen, there are dimensions/labels/rough guidelines for the features of the robot. It cannot be stressed enough that it is very important to ensure the dimensions of the robot work with the dimensions of game objects and structures as specified in the game manual. This helps ensure the CAD and integration goes smoothly later. Sketches do not need to be detailed renderings, but they should reflect the consensus strategy and include enough annotation to guide subteam work.

As can be seen in diagram 12.9, the gap between either side of the ball path is to be kept at 8.5". As the ball for the 2022 FRC Game is somewhat flexible and has a 9.5" diameter, a 1" compression distance makes sense. For the climb, multiple dimensions were tested until a set of values was found that allowed the combination of a single stage lift and pivot arm to traverse the monkey bars. This mechanism is described further in Subchapter 9.5.

Each team should determine the level of detail according to its own planning needs. As can be seen in this set of planning sketches, the structure for supporting the climb was not fully planned before CAD work on the climb began. Neither were key dimensions determined for the top view for the width and length of the indexer tower and the gaps between parts of the climbing mechanisms. Nor is the exact angle from the flywheel to hood labeled in the side view. Ideally this should have been done, which would have reduced the number of design iterations required.

At the same time, such dimensions may change later on as the design morphs and takes shape. Team 4817 decided to provide room for design changes and decided against labelling all of the key dimensions of the robot and only the ones that would ensure the mechanisms worked, (like figuring out if the lift and arm combination could solve the climb challenge and agreeing upon a 1" compression throughout the ball's path in the robot).

**Importance of Prototyping**

A good question to ask is how are key dimensions determined? Some can be planned out in a fully theoretical setting. Typically this is true for the mechanisms used in stacking type robots like claws, lifts, and arms. However, most key dimensions are best determined by practical testing. To some degree, teams can rely on recommendations from the FRC community. RI3D teams usually publish findings on what compression values to use for various mechanisms. However, the specific wheels and types of mechanisms that RI3D teams use may not be the same as the optimal design determined by a team. Often RI3D teams focus on making a complete robot as quickly as possible rather than the most optimal robot possible as this takes more time.

In that case, a team should conduct their own customized practical testing with the wheels, motors and parts they intend to use. There are a few ways to prototype mechanisms. One of the most efficient ways to prototype is to use V slots.

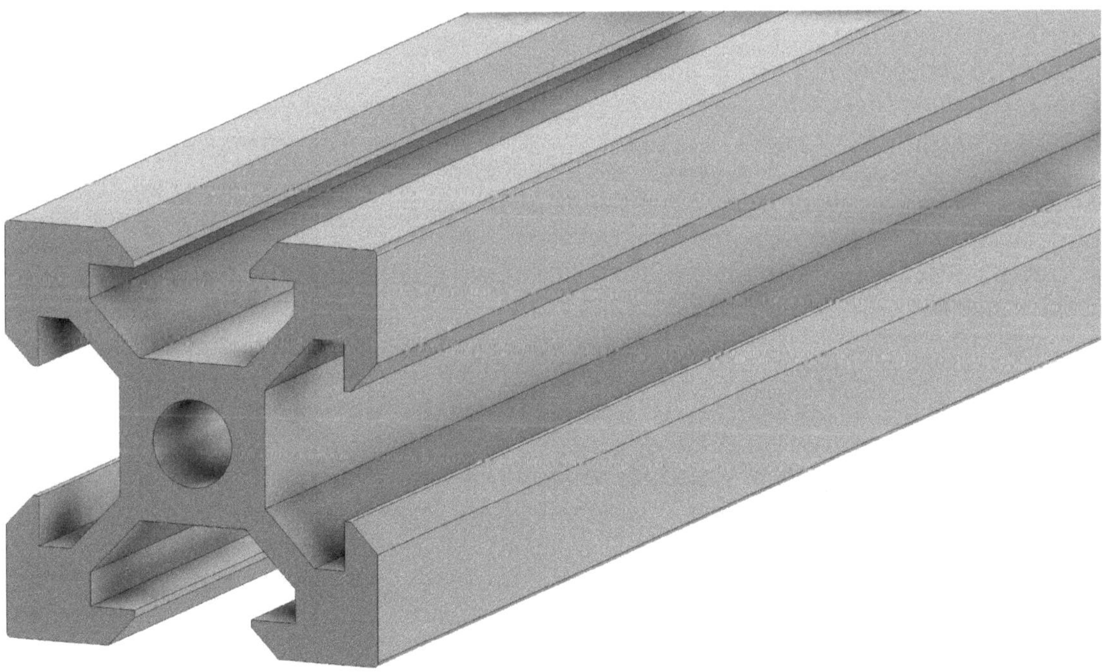

Diag. 12.10: V-Slot Slanted View; Ref. 40

A V-Slot allows plates to move along the V-Slot's length until bolts are fully tightened with T-Nuts to lock the plate in place. This allows for variable positioning along one axis. Combining 2 V-Slots together (attaching 1 V-Slot perpendicular to the other V-Slot) would then allow for 2 axes of positioning. (Combining 3 V-Slots would then allow for all 3 axes of positioning). With the right custom plates and V-Slot geometry, multiple compression distances can be tested by spinning a wheel with a cordless drill. This enables a team to determine the optimal compression distance for slower mechanisms. Distances can be measured with rulers, though calipers are more accurate.

For faster mechanisms, a custom gearbox utilizing an FRC motor is preferable to using a drill. Ideally the custom gearbox is designed to allow for multiple different gears to be swapped out to try different speeds. Theoretical gear ratio calculations for flywheels do not account for how compression affects shots. The way a specific wheel and game object interacts is dependent on the properties of the wheel and ball in real-world conditions.

Team 5987, a team that documents and shares their robot development process online, is an expert at prototyping. See their prototyping progress report for the 2024 Game at https://www.frc-handbook.com/ref/22.

## 12.6 The Handoff to the CAD Team

Now that a design has been selected, the CAD subteam gets started on modeling the design. Some teams go straight to building their robot based on the planning sketches described in Subchapter 12.5, however, there are real advantages to creating a CAD model first. It is not only an opportunity for team members to learn a valuable skill but also a way to conserve resources by planning out the robot before the build.

As highlighted in Subchapter 12.5, how well the robot actually functions for things like shooting and intaking is based on the key dimensions. Another way to prototype is to design and build a prototype mechanism as if it were the real mechanism, but to use a cheaper material instead. An example of this could be replacing a flywheel's Delrin plates with Plywood. The point of making these types of prototypes is to save money while confirming that the key dimensions are correct. These types of prototypes would be pretty similar to the final design, having the same shape and likely having all of the key parts of the design connected together. If a certain dimension is determined to not work optimally, revising the prototype is cheaper than revising a proper mechanism. Plywood is not durable enough of a material for FRC competitions, so teams should switch back to more durable materials once they are confident their key dimensions are accurate.

Once the robot design is adequately tested or the team feels pressed for time, the team moves onto full scale CAD production of the team's robot. In this phase, the CAD work is distributed within the CAD team. Every team does this differently. Team 4817 divides CAD work by assigning subassemblies to specific CAD members.

In larger teams, where there are more CAD members than subassemblies, multiple CAD members may be tasked to design the same subassemblies with the best design selected from the various options.

To ensure that the mechanical, electronics and programming subteams have enough time to work on the robot, it is essential that all CAD members complete their individual mechanisms on time.

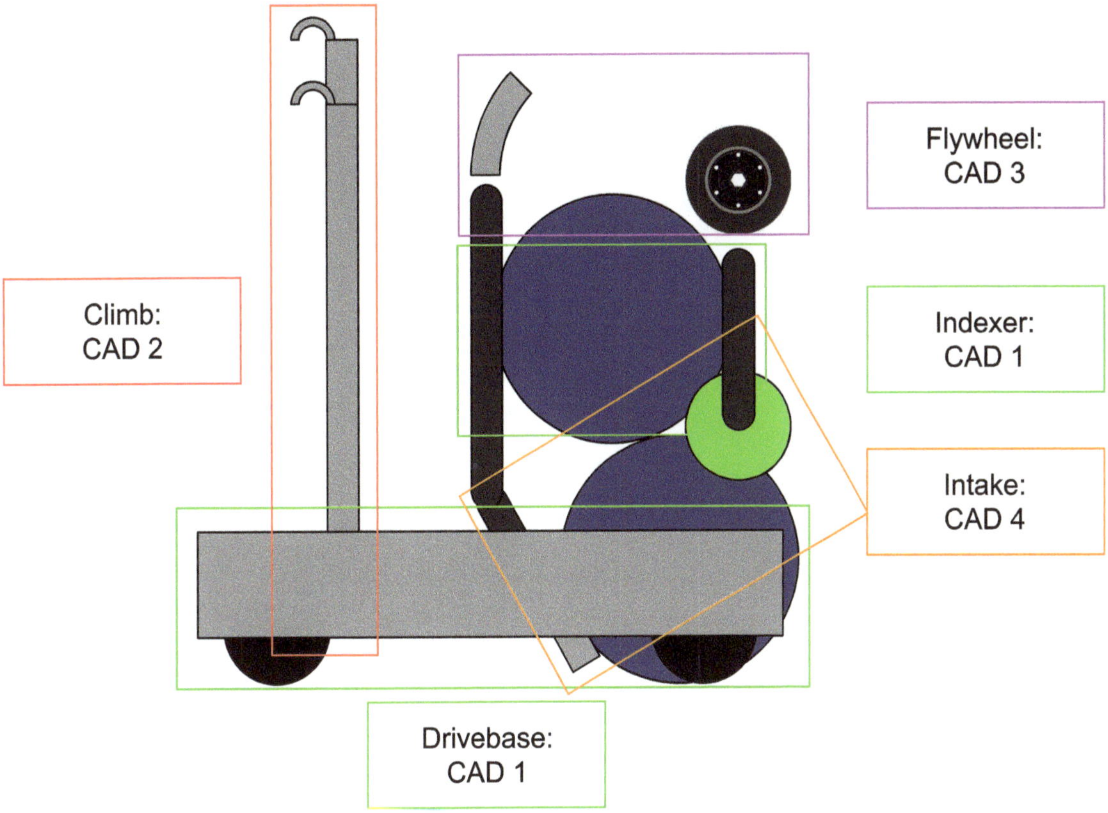


Diag. 12.11: Master Planning Sketch Distribution of Labor, 2022 FRC Game, Ref. 12

Once subassemblies are complete, they are combined into a master robot file. Ideally, this could be as simple as importing all of the subassembly files and the robot comes together perfectly. The likelihood that this happens will depend on how well the planning phase was executed. In situations where sections don't fit perfectly together, a few or perhaps all of the CAD subassemblies will need to be redone to make the integration work. Any number of conflicts can arise—dimensions may be off, paths may not align, or space constraints may be violated.

CAD work is not done until the full robot model is functional, within game constraints, and ready for fabrication. Precision here saves days of troubleshooting later.

## 12.7 A Macro Design Case Study

The process from initial discussion, to goal setting, strategy planning, robot planning, and finally developing a CAD model may seem a bit abstract. Even if a member understands the general process used for creating a robot design and all of the content knowledge in this book, it can still be challenging to figure out which set of mechanisms to choose to execute a game strategy. There are many mechanism options to choose from and different ways to combine mechanisms to complete the various objectives in games. There is not necessarily one set way a robot should be designed with a certain strategy in mind. This subchapter provides a real-world example of a team's journey from establishing a game strategy to the finalisation of a robot's macro design.

**2020 FRC Game: INFINITE RECHARGE**

In the 2020 FRC Game: INFINITE RECHARGE, robots collaborate to power up a city's Shield Generator by scoring Power Cells—7-inch yellow foam balls—into a goal called the Power Port. These lightweight balls are collected and launched by robots into three scoring zones for varying points. Robots can also manipulate a spinning Control Panel to activate Shield Generator stages and earn bonus points. In the Endgame, robots attempt to climb and balance on a swinging Shield Generator Switch for additional scoring. Refer to https://www.frc-handbook.com/past-frc-games for the full game menu.

**Game Strategy by Team 4817: Maximise Match Points**

We will use online Reference 11 for this case study. Team 4817 decided that their game strategy for Infinite Recharge would be to maximize match points. (See Subchapter 12.4 on Formulating a Game Strategy). They prioritized accurate shooting into the Inner Port, efficient Power Cell handling, Control Panel activation and coordinated climbing.

Given the strategy of maximizing match points, the robot had to be designed to score points as efficiently as possible. A way to think about this is that with the assumption that the robot's driving speed is consistent regardless of the mechanisms atop of the drivebase, the strategy should be to try to get the most number of points possible for the same number of actions. (This assumption is not necessarily 100% true especially for really optimized and professional teams, but generally the difference is not significant for average to even above average FRC teams.)

Reviewing diagrams 12.2 to 12.4 will provide insight to identifying the best scoring routes, intermediate destinations and end goals.

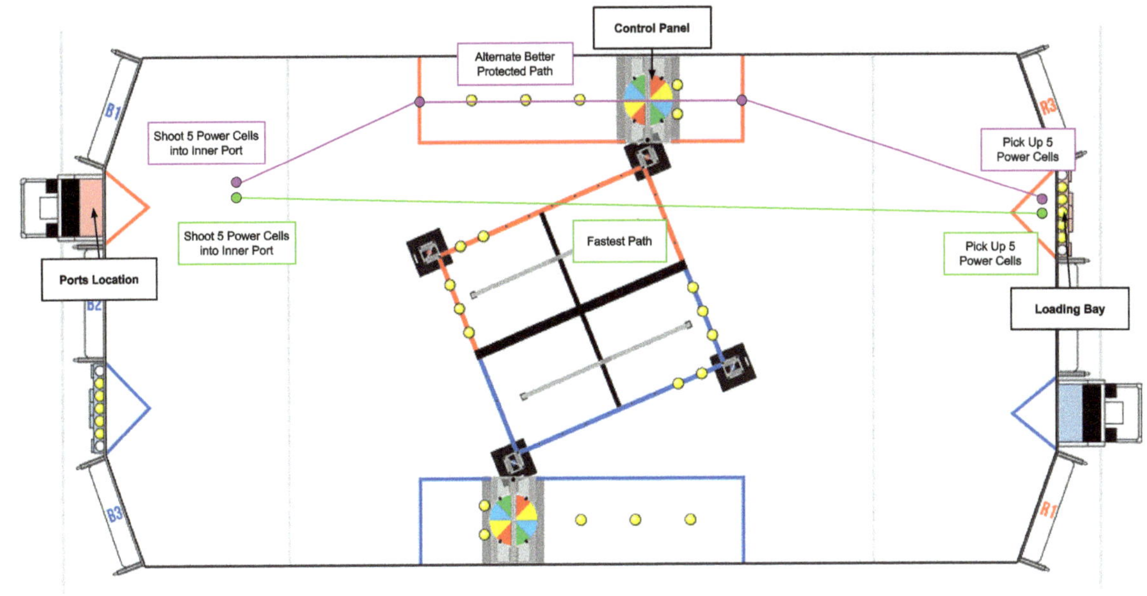


Diag. 12.12: Match Cycle Strategy, 2020 FRC Game

In each set of Power Cell possessions, the robot should aim for the Inner Port as that gives 3 points per Power Cell shot (compared to 2 for the Outer Port or 1 for the Bottom Port).

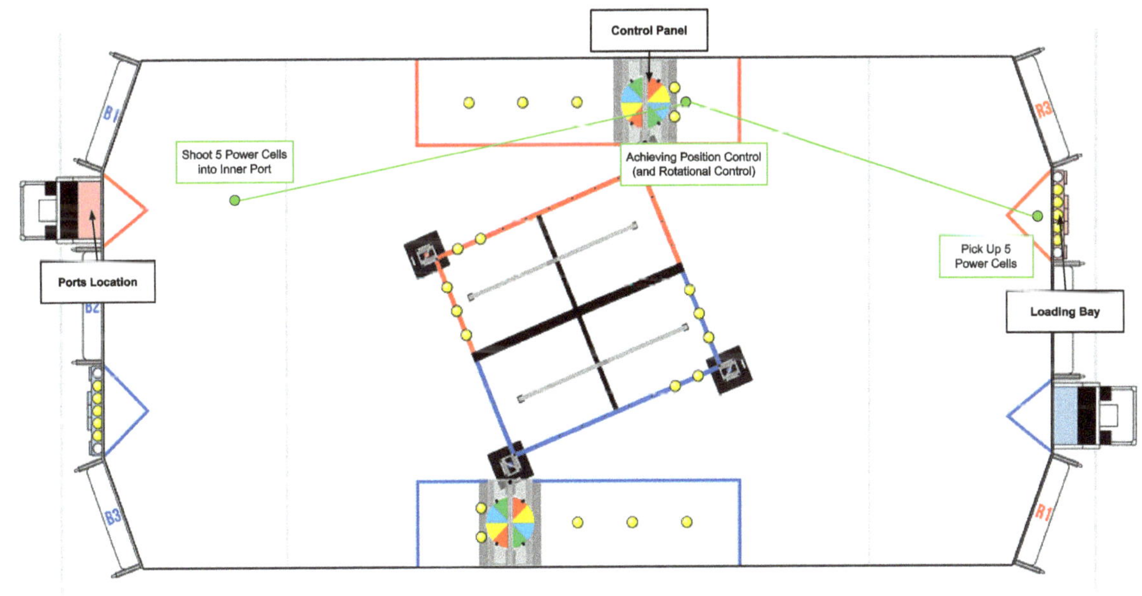


Diag. 12.13: Control Panel Cycle Strategy, 2020 FRC Game

Additionally, there are points to be scored for activating Rotational Control and Position Control at the Control Panel. Therefore, the robot needs an alternate path as highlighted in diagram 12.13.

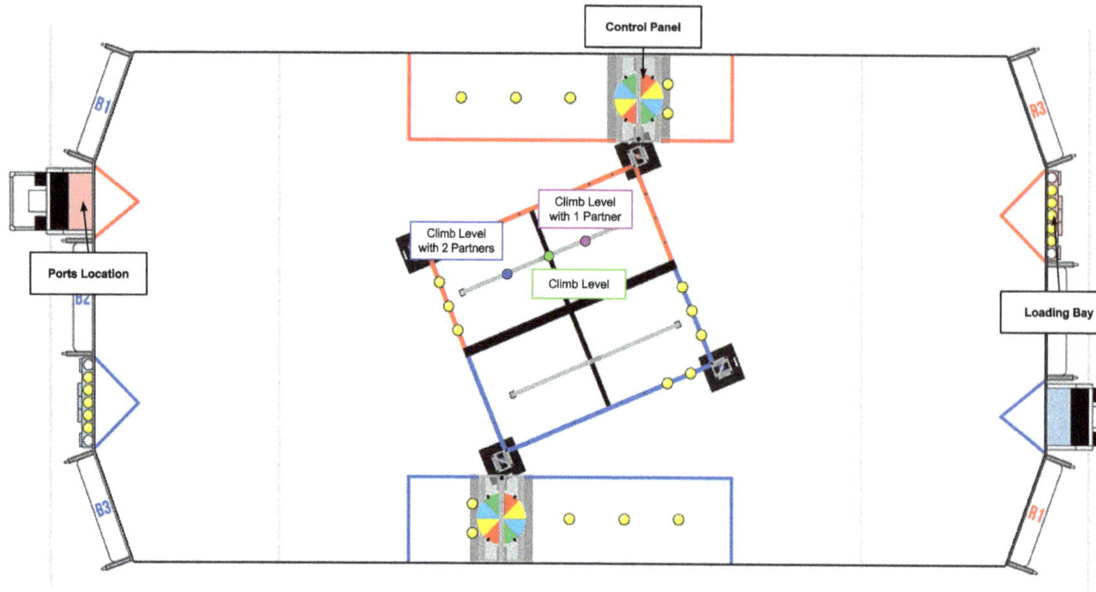


Diag. 12.14: Match End Strategy, 2020 FRC Game

Finally, there are climb points. Ideally, the robot is able to climb with the alliance partners onto the Generator Switch while keeping it level, but that is largely dependent on coordination, planning, and some level of luck (pairing with capable alliance partners) leading into each match. In terms of how a robot is designed to hang on a bar vs. hanging on the Generator Switch, the designs used are largely the same.

### Team Resource Limitations

In an ideal setting, a team could just plan out the best robot they can think of, but for most teams budget as well as available machinery does affect a team's design. A reality for Team 4817 that year was a very limited budget. The robot had to be designed in ways that made the best use of the electronics and hardware already in the team's inventory. Fortunately, Team 4817 had key machines to produce custom plates such as a laser cutter for Delrin and a CNC mill for Aluminium which allowed for greater design flexibility.

### Picking WCD for the Drivebase

The first step was to plan out the broad strokes of the robot beginning with the choice of a drivebase. Referencing the drivebase summary in Subchapter 2.11, the drivebase decision was relatively straightforward. WCDs were the only viable drivebase design given the team wanted something simple and relatively cheap. At the time, commercial swerve modules were not developed, so none of the complicated multidirectional drivebase designs were an option. (In modern FRC games, teams can buy off-the-shelf swerve modules which greatly reduces the complexity of using swerve designs. However, connercial swerve modules can be costly.) Recycling the previous year's WCD wheels and 6 wheel layout was done to save time. A 30" by 30" drivebase frame was chosen to maximize the robot's volume which in turn maximized the robot's storage volume. (For the 2020 FRC Game, the Robot's frame perimeter was capped at 120".)

### Shooting Robot vs Stacking Robot

The following flowchart was used to classify the optimum robot in terms of function—Stacking or Shooting. This helped the team to focus on a more limited set of mechanisms for the macro design.

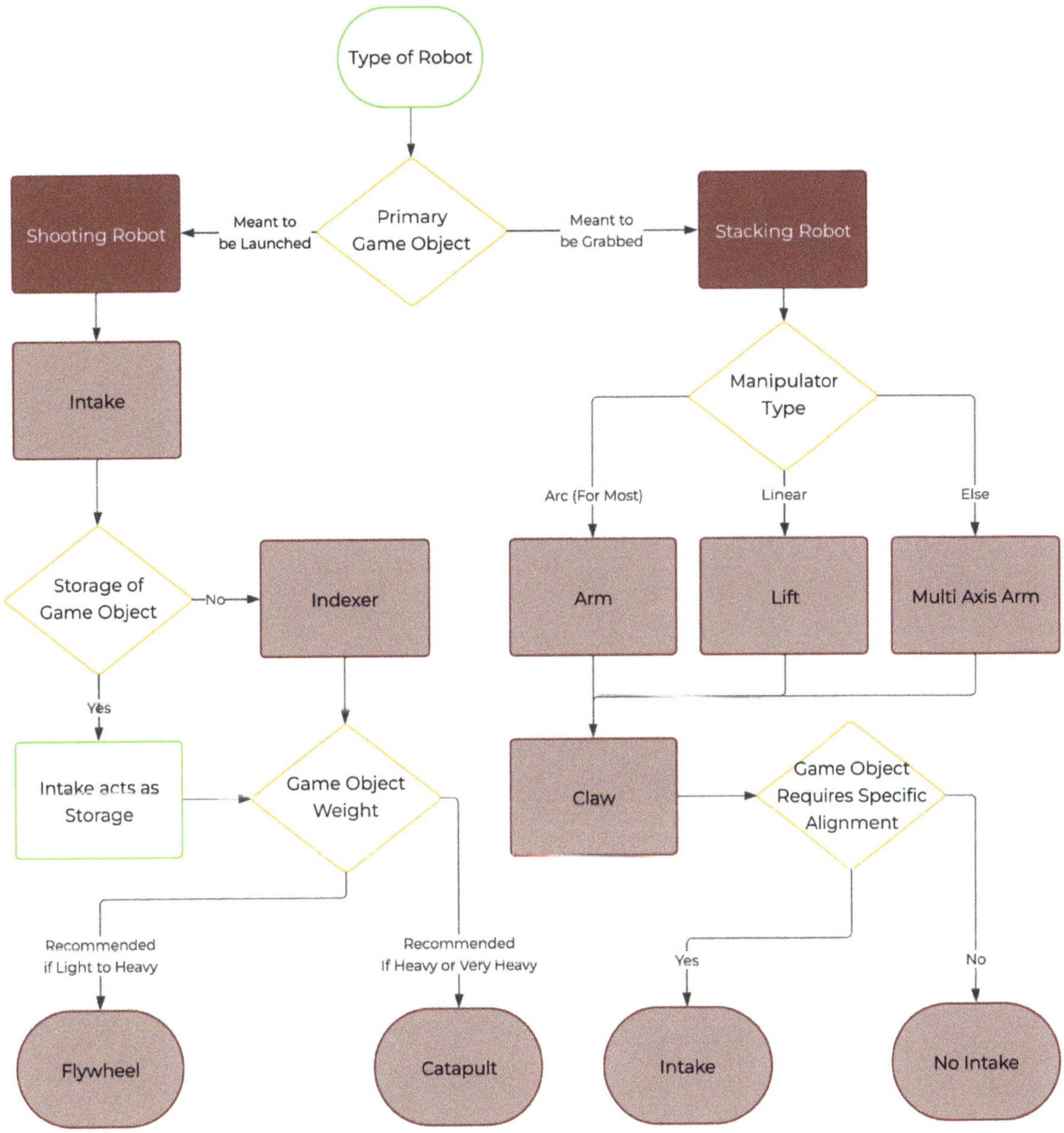


Diag. 12.15: Mechanism Flowchart

Given that Team 4817 wanted to score Power Cells in the Inner Port, as that scored the most points per Power Cell, the robot design had to be a shooting type robot. (The game rules prevented a stacking type robot from reaching high enough to deposit Power Cells in the Inner Port. Shooting the Power Cells would be faster anyways and was clearly what the game designers intended for teams to do with the Power Cells.)

## Ball Path For the Shooting Robot

Since this was a shooting robot, starting with the ball path through the robot was the logical next step. Given the Loading Bay's design and the general existence of Power Cells on the field, it would be optimal for the robot to be able to pick up Power Cells from both off the ground, and for Power Cells to be deposited into the top of the robot from the Loading Bay. This created a key dimension of less than 24" from the bottom of the robot to the top of the intake/storage area/indexer so that the loading bay could effectively deposit Power Cells into the robot. Given Power Cells were not simply being brought into the robot but also deposited into the robot, the intake/indexer needed to have an open top. The Power Cells being taken from the ground also needed to integrate with the ball path of Power Cells dropped into the robot from the loading bay. Since the Power Cells were lightweight, a flywheel launcher was preferred. However, the large size of the Power Cells meant that a hooded flywheel would not be very practical as that would necessitate the ball path going upward into the bottom of a hooded flywheel. Therefore, using a double vertical flywheel was chosen to save space.

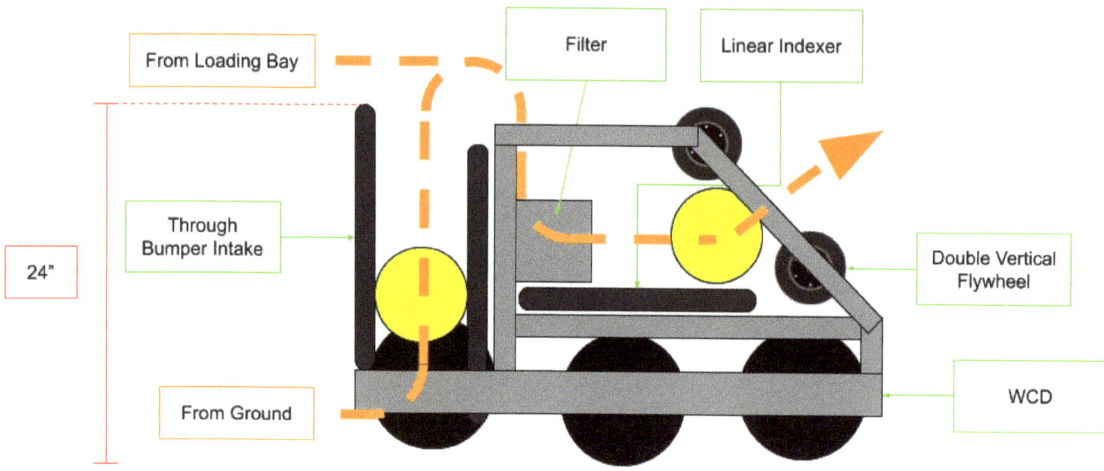


Diag. 12.16: Planning Sketch Side Initial Macro Design, Ref. 11

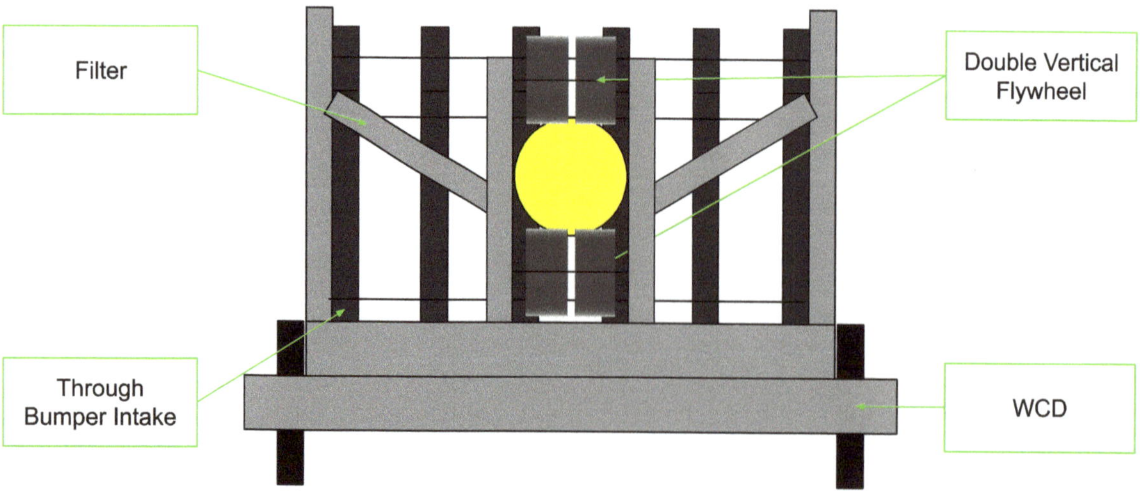


Diag. 12.17: Planning Sketch Front Initial Macro Design, Ref. 11

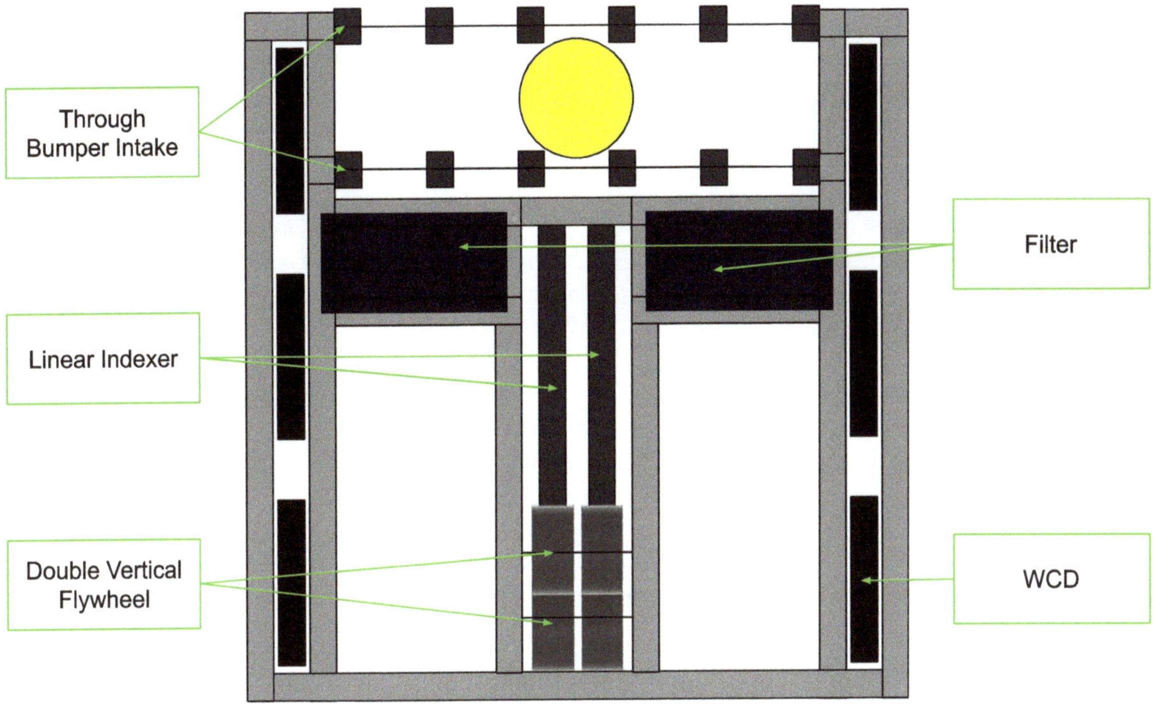


Diag. 12.18: Planning Sketch Front Initial Macro Design, Ref. 11

With this preliminary thought process, an initial macro design of the planning sketches were drawn up, per diagrams 12.16 to 12.18.

There are a few missing elements in this initial sketch—a climber, and a mechanism that interacts with the Game's Control Panel. This is a decent starting point, nonetheless. Next, the team considered its available resources in case that limited the design revisions in any way.

### Resource Allocation: Limelights & Motors

The first key resource to allocate was the LimeLight from the Team's inventory. LimeLights are a Vision Camera product that are able to act as the "eyes" of the robot. A LimeLight has vision like a human as it can detect color and depth, and can be integrated with advanced programming to make the robot "smart" and adaptable.

What could the LimeLight be used for? It could help with alignment to shoot into the Inner Port, detecting the orientation of the Control Panel, or identifying the positions of Power Cells on the ground.

The next question was: For each of these functions, what are the possible alternatives? For the Inner Port, the driver could just attempt to visually align the robot. For the Control Panel, the driver could also sight the Control Panel or use a cheaper color sensor (as budget allowed). For identifying Power Cells, again the driver could visually look for Power Cells on the field, but even more importantly, the benefit of using the LimeLight to pathfind to Power Cells on the ground was beyond the team's capabilities at the time. Moreover, the robot could be optimized for collecting Power Cells from the Loading Bay instead of picking them off the ground.

Given that landing Power Cells in the Inner Port had the most potential for scoring points (as Power Cells could be cycled many times during the game) and since utilizing the LimeLight for that purpose was also within the Team's programming capabilities, the LimeLight was allocated to the Flywheel.

The next key resource to allocate were the available motors. The team had 10 units of 775Pros, 4 units of CIMs, and a number of BAGs and Mini-CIMs. Even though the team had more than 16 motors, another important factor to keep in mind was that the legal power distribution hub at the time was the first generation PDP, which had 8 units of 40A ports, and 8 units of 30A ports. While modern brushless FRC motors are powerful enough that they can be used on any mechanism, the older brushed FRC motors were optimized for certain applications. Given this, 4 units of CIMs were allocated for the robot's drivebase, 6 units of 775Pros for the flywheels, 3 units of 775Pros on the intakes, 2 units of BAGs for the indexer, and the remaining assemblies would have to be covered by the remaining assortment of motors.

To summarize, the key questions to ask are: Where can x resources be used? What are the alternatives to not using x resource on y mechanism? If the best possible alternative option is used, does y mechanism become less effective and by how much? Then from there, it's a matter of weighing the pros and cons and making a strategic decision.

### Reconsideration of Robot Design: Incorporation of Pneumatics

With the motors allocated, a few apparent problems emerged. For one, using Mini-CIMs for a climbing mechanism would not be ideal given a climb's gearbox requires high torque to lift the entire robot off the ground. Up to 16 motor ports were allowed in Infinite Recharge. 15 motor ports were already allocated to power the intake, indexer, flywheel, and drivebase. That left 1 motor port to cover a climbing mechanism or a control panel mechanism. Choosing either mechanism meant forfeiting the ability to score points for the other, a trade-off that was suboptimal for overall performance. There was also a gear inventory issue for Team 4817. With a limited budget, buying additional gears—needed for climbing gearboxes—was not an option.

This is where content knowledge from Chapter 11 is useful as even though using such mechanisms complicates the design, (generally not a good thing), it allows for role compression with the same number of motors. The mechanism selected to save on motor count was the gear shifting strategy discussed in Subchapter 11.4, using the same gearbox to power either the drivebase or the climb. To accomplish this, pneumatics would have to be used in the robot to power the shifter in the gearbox. With the remaining motor slot, a small spinning wheel could be easily designed to interact with the Game's Control Panel.

Since multiple pneumatics can be powered by the same compressor, pneumatic control module, manifold (for solenoid valves) and air tanks, it made sense to consider pneumatics in areas other than the drivebase/climbing gearbox. A first thought was to change the through bumper intake to be an overbumper intake so that the storage of the robot could be expanded to the full 5 Power Cell limit for INFINITE RECHARGE. With the larger storage area, the indexer design needed to be switched to one with a greater capacity than a linear indexer that queues Power Cells in a straight line. Given that the Power Cells were being pushed to a flywheel positioned at the center edge of the robot, the logical indexer choice had to be a rotating indexer. Unfortunately, due to a limited budget, a full rotating indexer like the one shown in Reference 20 could not be built. Instead, a slanted plate and opposing rotating belts and wheels on the sides were implemented (see the first example in Subchapter 4.4 for details).

### Identifying Key Dimensions

A few key dimensions needed to be identified. Going back to diagram 12.13, one of the identified primary cycle paths involved going through the trench to access a protected zone in the field. The trench run necessitated the robot passing under the control panel, a key dimension of a retracted height of 28" was introduced. The next key dimension was for the climber. To reach the Switch at any angle, the full extended length of the single stage lift from the ground to the top had to be 79".

It was important for the climber to use a single stage lift to keep the robot design simple. However, the lift still needed sufficient reach to engage the Generator Switch. In order to adequately fit a tall enough single stage lift under the specified 28" retracted height, pneumatics could be used to angle the lift. That way, the lift could be pushed straight up going over the 28" retracted height only during the endgame. Designing the lift this way makes it an extendable arm—a type of multi-axis arm—with a rotational axis powered by a pneumatic actuator followed by a linear axis.

With all of that considered, a revised version of the planning sketches for the Final Macro Design was amended as follows.

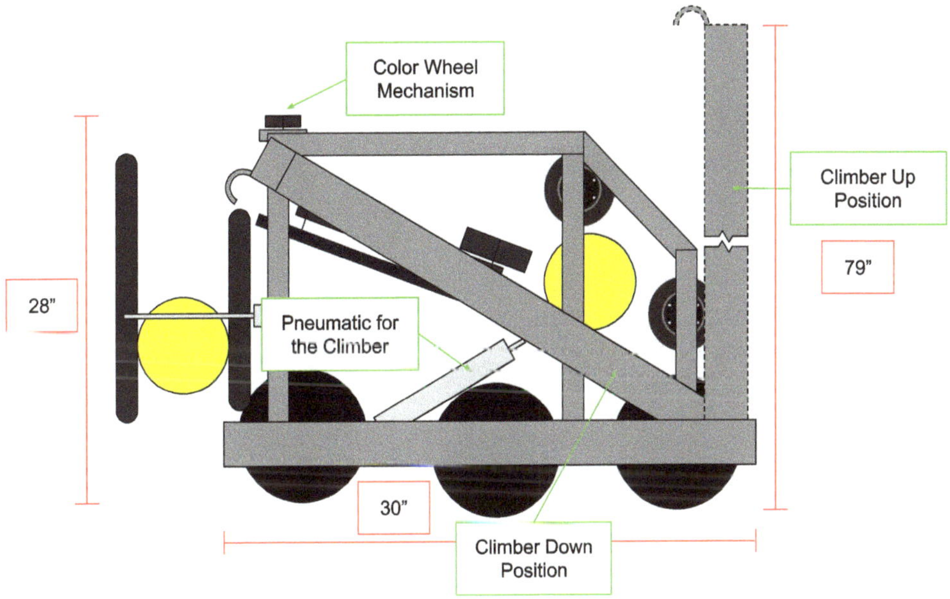


Diag. 12.19: Planning Sketch Side View Final Macro Design, Ref. 11

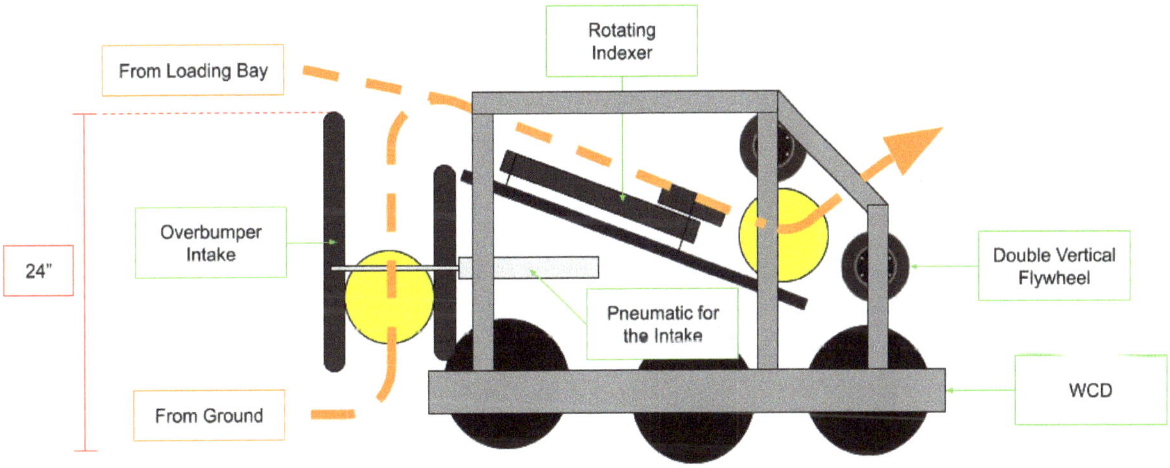


Diag. 12.20: Planning Sketch Ball Path Final Macro Design, Ref. 11

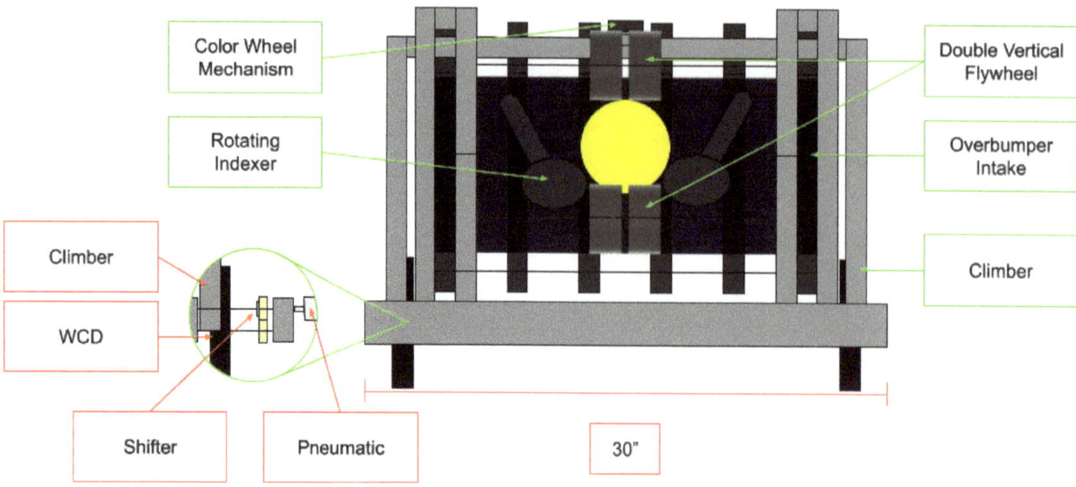


Diag. 12.21: Planning Sketch Front FInal Macro Design, Ref. 11

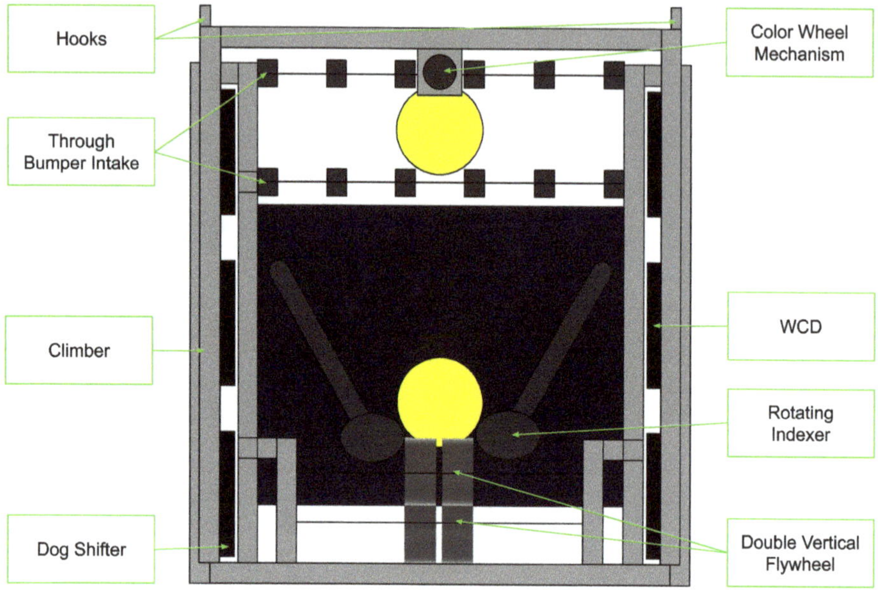


Diag. 12.22: Planning Sketch Top Final Macro Design, Ref. 11

This completed the macro design of Team 4817's robot for the 2020 game INFINITE RECHARGE.

## Flowchart of the Design Process

The flowchart in diagram 12.23 summarizes the thought process for this macro robot design.

On Kickoff Day, Team 4817 arrived at the Initial Macro Design within 3 hours of watching the Game Launch Video. Subsequently, team members ruminated on the initial design and arrived at the Final Macro Design a few days later. The mechanisms that were picked—over bumper intake, rotating indexer, double vertical flywheel, extendable arm & static claw—were incorporated and integrated by the CAD Team.

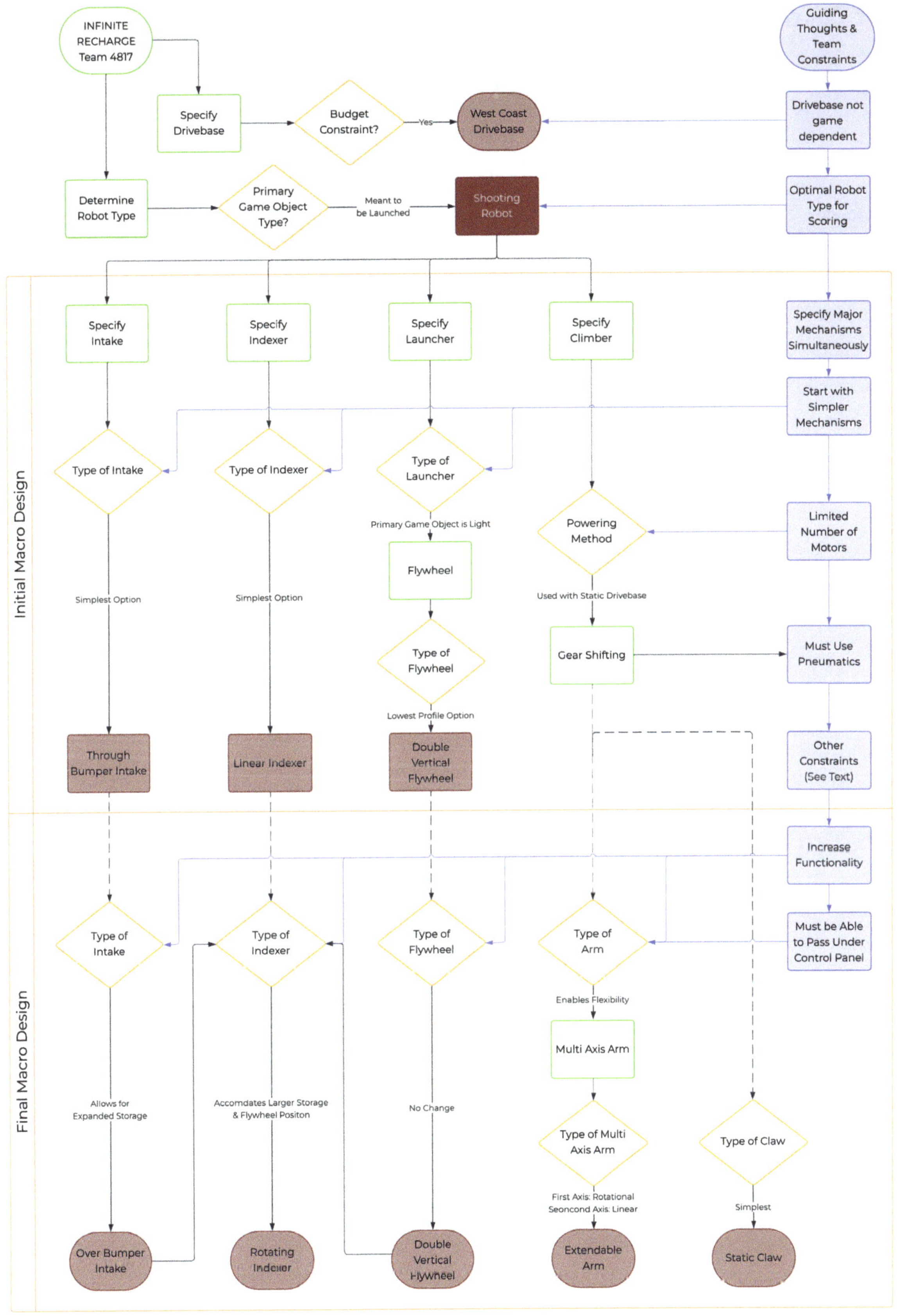


Diag. 12.23: Reference 11 Design Flowchart

## 12.8 Closing Reflection

The Kickoff process is more than a meeting; it is the foundation of the season's success. It combines creative vision, structured decision-making, and strategic realism. When each step is approached with thoughtfulness and rigor, your team sets itself on a trajectory not just to compete—but to excel.

It is hoped that this book has helped guide your thinking, sharpen your planning, and prepare your team for what lies ahead.

Begin your season with clarity. Stay true to your strategic intent. Build with foresight. That is how great robots, and great teams, are made.

Wishing you a fun and rewarding season of learning, innovation and growth.

## SHARE YOUR EXPERIENCE

If this book has helped you in your FRC journey, please consider leaving a review on Amazon and/or Goodreads.

Your feedback helps more students and mentors discover this resource and contributes to a stronger, more connected FRC community.

Amazon Review

Goodreads Review

Thank you for taking the time to support this work. Your review may help another team take greater ownership of their design process as a team—and bring their next great robot to a new level.

www.ingramcontent.com/pod-product-compliance
Lightning Source LLC
LaVergne TN
LVHW071412210926
848486LV00004B/260

* 9 7 9 8 9 9 9 7 8 2 8 1 6 *